Science Education
as/for **Sociopolitical Action**

Studies in the
Postmodern Theory of Education

Joe L. Kincheloe and Shirley R. Steinberg
General Editors

Vol. 210

PETER LANG
New York • Washington, D.C./Baltimore • Bern
Frankfurt am Main • Berlin • Brussels • Vienna • Oxford

Science Education
as/for
Sociopolitical Action

EDITED BY
Wolff-Michael Roth
and Jacques Désautels

PETER LANG
New York • Washington, D.C./Baltimore • Bern
Frankfurt am Main • Berlin • Brussels • Vienna • Oxford

Library of Congress Cataloging-in-Publication Data

Science education as/for sociopolitical action /
[edited by] Wolff-Michael Roth and Jacques Désautels.
p. cm. — (Counterpoints; vol. 210)
Includes bibliographical references and index.
1. Science—Study and teaching. 2. Postmodernism and education. 3. Science and
state—Citizen participation. 4. Technology and state—Citizen participation. I. Roth,
Wolff-Michael. II. Désautels, Jacques. III. Counterpoints (New York, N.Y.); vol. 210.
Q181 .S3687   507'.1—dc21   2001038380
ISBN 0-8204-5696-9
ISSN 1058-1634

Die Deutsche Bibliothek-CIP-Einheitsaufnahme

Science education as/for sociopolitical action /
ed. by: Wolff-Michael Roth and Jacques Désautels.
–New York; Washington, D.C./Baltimore; Bern;
Frankfurt am Main; Berlin; Brussels; Vienna; Oxford: Lang.
(Counterpoints; Vol. 210)
ISBN 0-8204-5696-9

Cover design by Dutton & Sherman Design.

The paper in this book meets the guidelines for permanence and durability
of the Committee on Production Guidelines for Book Longevity
of the Council of Library Resources.

© 2002 Peter Lang Publishing, Inc., New York

Printed in the United States of America

Contents

1

Science Education as/for Sociopolitical Action: Charting the Landscape

Wolff-Michael Roth & Jacques Désautels

> Any educational system is a political way to maintain or to modify the appropriation of discourses, with the knowledge and the power they carry with them. (Foucault, 1971, p. 46, our translation)

Disasters attributed to human agency continuously make the news. Three Mile Island, Chernobyl, *Exxon Valdez*, and Erika are but a few names associated with environmental disasters related to the energy sector. Scientific or technological reasons are cited when it comes to attribute blame, especially because such disasters not only threaten people's lives but also the livelihood of those within the larger sphere of influence of the event. The involvement of citizens to set policy for technoscientific developments or to assess controversial technology is one of the (more democratic) ways to improve decision making about science and technology. Thus, for example, consensus conferences involve citizens as a jury deliberating on issues of high technoscientific content and moment. The general aim of consensus conferences is to expand access beyond traditional (elitist) scientific and technological "experts," to increase public understanding of science, and enhance the democracy. There exist other forms of participation that allow citizens to contribute, in different degrees, to the setting of policy relating to technoscientific issues. Among these forms are referenda, public hearings/inquiries, citizens' panels, public advisory committees, and focus groups. Of course, because all of these mechanisms extend the range of participants with respect to past practices, this amounts to a democratization of decision making. An increasing democratization of technoscientific decision making and technology assessment perhaps requires greater competence for participating in technoscientific discourse. Yet there are problems.

On the one hand, there are growing calls for greater public involvement in drawing up policies that pertain to science and technology in our society, in line with and reflecting democratic ideals and for enhancing the trust in regulators and transparency of regulatory systems (Rowe & Frewer, 2000). On the other hand, there appears to be a lack of input from the general citizen who is not an expert, not a representative of an interest group, and an immediate stakeholder (Guston, 1999). From a science education perspective, the problem may be framed in terms of a democratization that would require greater "scientific literacy." Thus, scientific and technological literacy has become a rallying cry in recent reform movements (AAAS, 1989; NRC, 1996). Yet these developments appear as just another twist to continued reform, which, for science education in the United States, began with the shock waves rippling through society, caused by *Sputnik* and an apparent Soviet advantage in space science and technology. That is, we are less than convinced because the reform does not touch the deep problems that surface when we look at the way science is related to other aspects of modern life.

We are quite disheartened with past attempts of reform and think that the fundamental conditions of schooling and of what schools are for have not changed. For instance, as soon as we look at the operationalization of the concept of "scientific literacy," we find a reactivation of past schemas: scientific content that has to be acquired by students from K to 12 (Eisenhart, Finkel, & Marion, 1996). The problem of science education is then framed in terms of the question, "How do we create the conditions for students to experience and acquire the scientists' way of knowing?" Science educators frame *this* question in *this* way as though the formation of citizens who can act as critical social actors and engage in sociotechnical and environmental controversies, inevitably follows from such educational endeavor. In other words the revolutionary potential of postmodern discourse on scientific literacy is diluted in the pedagogical propositions brought forth by curriculum "experts." Thus, the only difference we see that (radical) constructivist referents made to science education as practiced is to give a greater focus on learning processes. But the basic message to students and teachers remained: "Learn/teach the science of scientists." Somehow scientism is back in the foreground dressed up in different clothing. In this, even the most recent reforms informed by constructivist rhetoric are committed to symbolic violence mostly disregarding common sense, vernacular, and indigenous forms of knowing. However, scientists' science is not the only goal that one might set for school curriculum. As the title of our book indicates, we frame this goal largely in terms of participation in sociopolitical action.

Science Education for Social Action . . . !

Perusing some of the recent literature on the purposes of science education, we notice the equivalent of a paradigmatic shift in the conception of what should be the ends or aims of science education. For a long time the teaching of the scientific disciplines was a taken-for-granted necessity. This is not the case anymore. Some curriculum theorists propose that scientific literacy for everyone should become the ultimate aim of science education (e.g., Désautels, 1998; Eisenhart, Finkel, & Marion, 1996; McGinn & Roth, 1999).

These authors acknowledge that science education as it was practiced did not succeed as an enterprise and that we should give more consideration to the work of scholars advocating an STS perspective on science education. This literature often draws on the work in the field of the social studies of science which changed in important ways our representations of the sciences and also, in a lesser extent, on the work on the public understanding of science. Common to the different approaches of rethinking science education are the ways in which ordinary citizens are conceptualized. Rather than conceiving of them in a deficit mode, ordinary citizens skillfully appropriate scientific knowledge for their own purposes by taking into account the recent development of our societies towards what Beck and Giddens claim to be manufactured "risk societies." Thus, there is more and more evidence that citizen groups have not only substantial influence on the specific research questions scientists pursue (e.g., Rabeharisoa & Callon, 1999), but also on the methods and protocols that scientists employ in the pursuit of answers (e.g., Epstein, 1997).

Therefore, instead of making the focus on the teaching and learning of disciplinary knowledge, the empowerment of citizens to become critical social actors is now conceived as the goal towards which we should strive. Certainly there are science educators who will not agree with this outlook. There are still debates about the nature of scientific literacy. But we feel that time has come to take position and elaborate on these matters and provide examples of scientific literacy in praxis. For example, whereas traditional science education viewed scientific literacy in terms of the acquisition of scientists' science, some recent articles reframe scientific literacy in terms of knowing what (scientific) resources to draw on and where to find them (Fourez, 1997; Roth & McGinn, 1998).

There are at least two different routes to scientific literacy. One is that we focus on the teacher who, like Paulo Freire, engages with learners in such a way as to assist them in transforming their reality. Angie Barton (see Chapter 8) engages children in homeless shelters in activities—which older colleagues often do not consider science—but which have tremendous potential of reaching children and providing them with a place from which to transform themselves and their lifeworlds. Ken Tobin's work and thoughts on street science appear to us to go into the same direction—though the problems he is facing (see Chapter 6) may themselves point to a problematic approach to the original problem. On the other hand, we might think of science education as a by-product as our students engage in social action. Here, the focus is much more on learners who transform the community in which they live through their actions. Jacques Désautels has been working with a teacher collective that gets its students into situations where their actions transform some situation: students learn about human biology as they produce leaflets and conduct one-on-one campaigns around reproductive issues among working class women (e.g., Collectif chimie Cégep Limoilou, 1998). In this way, science is no longer separate from social and political action. In a similar way, Stuart Lee is part of an activist group that attempts to transform the community in terms of its water-related practices that impact on the local watershed. Here again, science is part and parcel of a larger network of activities that are social and political.

Fourez (1997) also has done interesting work on what it means to become scientifically literate (see Chapter 3 by Lee and Roth, an elaboration on and examples of Fourez' criteria). No longer does it mean to memorize or even become competent in scientists' science, but it means to be able to draw on a large number of resources that might be categorized as science and science process. In the same vein, Roger Cross and Ronald Price have argued in the past that science teaching should have social responsibility as its goal (e.g., Cross & Price, 1992). All of these ideas run cross current or counter to the traditional way of thinking about science, science teaching, and reform. We thought that the time is ripe for this book, a manifesto, in which we different authors have the space to develop our ideas about science education as/for social action.

Science Education as Social Action . . . !

At a macro level of analysis one can legitimately think that the main function of science education in our school systems is to reproduce the social hierarchy of knowledge and the powers that are related to them (Foucault, power/knowledge). For example, it is relatively easy to show that science education is a sub-system of science. This sub-system, through a re-production cycle, which begins with kindergarten and goes through university-level education, produces teachers who, without malice, teach dogmas in a pattern that form more than one link with the teaching of religion.[1] For instance, the nature of science as producing truth, as it is taught in most school science classrooms, is equivalent to the

---

1. Fuller (1997) points out that there are considerable structural similarities between traditional religions and science.

dogma of the Immaculate Conception of the Catholic Church. Thus, scientific research is said to yield truth, whereas it is only by means of social influences that this truth can become tainted. More so, some researchers claim that one can interpret what goes on in school laboratories as a kind of ritualistic activity (e.g., Lessard, 1989).

One can also analyze science education as "social action" in adopting a micro-sociological perspective. For instance, Roth and McGinn (1998) use an actor network theory to analyze the local processes (the production of inscriptions in particular) that, through their circulation, distribute power in the community all the while deleting, that is making invisible, the lives, work and voices of those being inscribed. Grades and grading practices constitute a paradigmatic case of this distribution of power/knowledge within an educational network.

Traditional models of public involvement in science and technology policy are based on the view that decisions should be left in the hands of experts and scientists (Rowe & Frewer, 2000). Often, these models are based on a deficit view in which the inadequacies of the general public with respect to science are said to limit its capacity to be effectively involved in complex decisions. Apart from ignorance, other factors may diminish the potential for participation such as attitudes, beliefs, and motivation.

Perhaps the most pervasive argument for public involvement is that value judgments are made at all stages of policy development. This includes judgments such as deciding which risks to evaluate (Rowe & Frewer, 2000), and it includes judgments about whether the potential dangers from treatment not given outweigh the certainty attributed to scientific knowledge derived from double-blind studies and the administration of placebos. For example, a major inroad of AIDS activists to science policy and even the nature of science came through their insistence on being administered unknown but potentially successful treatments rather than placebos that are known to have no effect (Epstein, 1995). The insistence of the AIDS activists on receiving drugs rather than placebos, and their participation in the public dialogue about AIDS and treatment, led to changes in the way drugs are being tested (Epstein, 1997). In effect, then, although AIDS activists are far from being medical or pharmaceutical experts, they were able to change the protocol scientists had to use to test new drugs.

Situated cognition theories, particularly those that focus on the affordance of learning by participating in communities of practice (e.g., Lave & Wenger, 1991), highlight the tremendous benefit to learning by participating from early on in day-to-day activities of the domain one is to acquire. Thus, we learn most of what we know (including common sense) not by explicit acquisition of propositions but by participating in everyday life. The power of this learning can be taken from the fact that we learn our mother tongues not by acquiring a grammar first and then speaking in correct sentences. Many adults do not know formal grammar and yet have learned to speak properly and are able to distinguish well-formed sentences from those that are not. There are many reports from traditional societies that show how learning accrues through participation in the everyday activities of some domain of knowledge. For example, Vai tailors learn their trade by participating in increasing ways in the daily production of clothes (Lave, 1977), and Mexican Indian (Maya) midwives acquire what they know through a life-long participation in different aspects of the daily affairs related to pregnancy, story-telling, birthing, and so forth (Jordan, 1989).

What is particular about these cases is that the learner does not participate in fake activities, designed to provide them with knowledge that they are then to apply in a different context. Rather, in these situations, individual learning is equivalent to their changing participation in ongoing activities (Lave, 1993).

These examples may encourage us to rethink the ways in which we conceive of science curriculum. Thus, rather than thinking of school science in terms of scientists' science, we may think of it in terms of participation in public affairs related to science and technology policy. Furthermore, this participation does not have to be delayed to some future point in time. Rather, as shown by many authors in this book, children and young adults are perfectly able to participate in science-related activities that contribute to a larger good.

Of course, such early participation may have political implications. For example, one of the students in the school in which Roth works conducted a study on the coliform levels in a creek. He sought help at the university to analyze his water samples. He found a particularly high level of coliform bacteria in the creek below a chicken farm. Encouraged by a sense of success that he understood how to sample and analyze water for coliform bacteria, he wanted to collect further samples. However, he found himself barred from accessing the creek again in the vicinity of the chicken farm, whose owner was dismayed and possibly afraid of the possible impact of the student's findings. This potential impact was especially salient given that an environmental activist group (see Lee/Roth chapter) is very active in making the community aware of the impact various practices have on the health of the creek.

Public participation in policy making in science and technology is necessary if society is to reflect democratic ideals and to enhance trust in and transparency of regulatory systems (Rowe & Frewer, 2000). One of the main questions for science educators has to be how we can prepare students for such participation.

Public understanding of science is often evaluated in terms of standard scientific facts, methods, and attitudes. However, there are no reasons why these dimensions are better than other sets of dimensions. For example, institutional knowledge of science and attitudes to the nature of science are alternative dimensions in terms of which public understanding of science can be measured (Bauer, Petkova, & Boyadjieva, 2000). These authors argue that if ordinary citizens had a high degree of institutional knowledge of science, they would also be able to trust science as an institution.

Charting the Landscape

There is more than one way to draw a map of an unknown territory, and the ordering of the chapters in this book represents one of the many possible pathways that can be explored in order to establish its contours. We thought that it was wise before we started on our journey to first consult an elder (Jenkins, Chapter 2) who could, given his long experience in the field, make connections with past experiences, identify potential obstacles and raise fundamental issues. Jenkins therefore is our mapmaker of a terrain that the other authors explore in greater detail. We, the editors, serve as the reader's guide both here and at the beginning of each of the three chapter clusters. (We provide a more traditional "abstract" for each chapter in the appendix so that readers can construct their own reading itinerary through the book.)

Jenkins introduces us to the idea that the connections between science teaching and sociopolitical and practical issues of everyday life have a long history. He also suggests that most of the pedagogical solutions that were experienced had marginal effects regarding the transformation of the traditional school science education. It seems that the ideology underlying conventional school science was grounded in a positivistic epistemology. This epistemology constitutes a major obstacle in the design of any project oriented towards a form of science education for social action. In fact, reformers of all venues are reminded that in any society, schooling is the main social process through which particular discourses are culturally reproduced and thereby establish their domination in the public sphere. Therefore, the central question of power/knowledge as framed by Foucault has to be in the back of our minds all along our journey through the landscape traced by the chapters in this book. More so, our mapmaker reminds us that any attempts to promote scientific literacy as goal for science education should take into account the complexities inherent in this social endeavor.

First there are socio-epistemological complexities. We have to acknowledge that the classical idealistic representation of the sciences as intellectual feats enabling human beings to decipher the secrets and the order of nature has been fractured. Anthropologists and sociologists of science have called into question the image of science as a way of accessing truth-seeking practices. In contrast, the newly emerging image stresses the local and contingent nature of scientific practices. On a macro level, some sociologists articulate science and technology in terms of the interests of those who gain from advancements in the military domain or of the exploitative practices of large transnational companies. Science and technology are therefore deeply integrated to and part of more inclusive sociopolitical practices. More so, in a society that

continuously manufactures risks—such as the disasters we referred to earlier—*society* has become the laboratory. The presently
uncontrolled and proliferating distribution of genetically modified
organisms in the human food chain is but one example. To maintain a minimum of democratic life, citizens must therefore have
their say in what was traditionally considered as the privileged
domain of scientists that is scientific policy. Traditional science
education does not prepare students to become competent actors
in their society. (Only small portions of students take science until
their senior years. Many adults claim that they do not understand
science and hated it as school subject.) But, other complexities
have to be considered if such a goal, often labeled scientific literacy, is to be pursued.

Engaging students in the solution of meaningful social problems and having them enact scientific practices in the everyday
world rather than in the confines of well-controlled school laboratories is a demanding task both for the teachers and the students.
It involves students in a form of practical reasoning while tackling
the messiness of things that do not necessarily obey the classical
well-formed mathematical relations between variables. In the everyday world of action and decision making, knowledge is most of
the time uncertain and contentious. Students might even discover
that in certain circumstances scientific knowledge is irrelevant. In
other words, involving students in meaningful social actions such
as in environmental controversies or problems is a risky endeavor.
But it is a risky endeavor that is promising as illustrated by some
of the case studies that our mapmaker sketched out. Science education for social action is certainly a pedagogical possibility, one
that he conceives as worthy since it is about the empowerment of
students as citizens. But we should not be naive about such matters since political education is a very touchy subject indeed, and
the "Science Wars" have shown that the scientist community may
show a lot of hostility towards a project grounded in a more rela-
tivist epistemology.

With these deliberations in mind, we organized the material of
the book into three clusters, each one echoing and completing our
mapmaker's comments on science education for social action. In
the first cluster, the reader will find two chapters by Lee and Roth

where they report on their research regarding an environmental project in a small coastal community in British Columbia, Canada. Different actors are involved in the environmental issue including students, parents, activists, farmers, and First Nations native people. In both of their chapters, the authors take a strong stance on what it means to be scientifically literate: the competence to participate in the conversations and debates about significant social problems. The authors go on to show how one cannot dissociate knowledge production and participation in the community. They demonstrate that the students involved in the project (building a riffle in a brook, studying the quality of the water) could create a significant amount of genuine knowledge and become knowledge brokers within the community. In short, they support a piece of evidence for our mapmaker's claim that science education for social action is no utopia but radically transforms existing practices. With these authors we may say that science education for social action offers a realistic solution to the quest for scientific literacy for all without its usual scientistic and dogmatic overtones.

The second cluster of chapters charts territory that echoes our mapmaker's sketch in more than one way. Cross and Price (Chapter 5) have been known for a long time as advocates of a critical position in science education. From their point of view, today's schools must, more than ever, provide a type of science education that will help all students, especially future scientists, to become advocates of a socially responsible science. The proposal by Cross and Price retains some of the classic positions held in the science education community whereby the teaching of controversial subjects (nutrition and health, manufactured food, sustainable agriculture, genetically engineered food production, etc.) should not interfere with the learning of the fundamental concepts of the sciences. However the authors insist that these concepts should not be mere rhetoric but rather presented as the products of social processes involving scientists in debates and controversies about fact production. More so, they suggest that students be involved themselves in debates and discussions about the political, ideological and ethical dimensions of the different social uses of scientific knowledge. Their position resonates with Jenkins' comments,

in particular when they point to the fact that schooling is a powerful relay for the ideology of the scientific community. Cross and
Price resonate with our mapmaker even more in stressing the complexities involved in the decision-making processes.

The remaining chapters in this second cluster draw on the respective authors' experiences from the field. These experiences
allow raising questions about the pertinence of saving one part or
the other of the classical science curriculum in the context of a science education for social action. For instance, Tobin's (Chapter 6)
experience in teaching proper science to disadvantaged African
American students in a Philadelphia high school is a telling one.
Even though he made astonishing efforts to draw on the students'
neighborhood experiences to motivate them in learning science, he
experienced from their part the greatest resistance one can imagine. They were not the least interested in studying garbage dispersion in their environment (which could have led to the study of
health and nutrition) or any of the *Street Science* themes he proposed. Reflecting on this particular experience, Tobin concluded
that the central issue was not one of curricular foci but one of
finding a way to build a community constituted in rapport and
mutual respect of all participants. We learn from this study that
imposing a socially relevant curriculum because we *think* it addresses the needs of students can go terribly awry.

The authors of the subsequent three chapters seem to have
overheard Tobin's discourse. The centrality of the community and
participation of its members lies in their participation in defining
the relevant matters at hand and in defining what are worthwhile
science education experiences. What counts as legitimate knowledge and how it is put to use (in order to accomplish some social
action) is not imposed from above. Rather, what counts as legitimate knowledge is the result of deliberations and negotiations between the actors. This is the case whether curriculum is defined in
the context of a First Nation community (Aikenhead), providing
science experiences to "at-risk" children who live in shelters (Barton & Osborne), or the use of scientific discursive resources by
women militating either for or against abortion (Lawrence &
Eisenhart). After reading these chapters one is ready to reflect on
our mapmaker's final question: What is school science for?

The last cluster of chapters is organized around what may seem to be an esoteric theme, that is, epistemological practices. But all authors, including our mapmaker, either explicitly or implicitly visit the topic or call into question the positivist version of the production of scientific knowledge. In this cluster of chapters, the authors go one step further. First, they claim that one cannot separate matters of epistemology from politics, ethics or aesthetics. Second, the authors also claim that epistemology does not constitute an esoteric practice. They demonstrate that even elementary school children can engage in such practices and finally, that all knowledge whether scientific or common, including their own, can be potentially questioned if an authentic participatory and democratic debate is to be fostered in our societies. In short, the three chapters foreground the necessity that all forms of knowledge need to be open for interrogation in order to create a space for a more democratic lifeform.

Launching the discussion, Larochelle (Chapter 10) shows that schoolteachers' imposition and re-institutionalizing of scientific categories at the expense of other ways of knowing can have very perverse effects on children's spontaneous epistemology. Under these conditions, children do not have the opportunity to understand that knowledge, scientific or personal, is the result of social processes. When children are forced to learn what knowledge counts without being able to interrogate it, they are literally disabled to develop the competence for questioning of any form of knowledge. They are therefore stunted in their growth in interrogating the ontological value of scientific knowledge and in participating in reflective ways in debates that have both social and ethical implications.

Désautels, Fleury, and Garrison (Chapter 11) pursue a similar theme. They show that elementary school children enact epistemological practice through their discussions. They also illustrate how adolescents can question the value of their school science education experience and the infallibility of scientific knowledge and, therefore, develop a relation to knowledge that socially empowers them. Finally, they emphasize how the aesthetic dimension of the students' ways of knowing should be taken into account, understanding how they make sense of their school science experience. In the last chapter, Roth and Bowen make an argument for the teaching of epistemology by drawing on their research in middle and high school science and among practicing scientists. Their analyses show how an appropriately designed learning environment allows students to develop a diversity of epistemological stances and to argue about the status of language in the production of scientific knowledge. This competence in discussing such problematic matters can be easily extended to interrogate their everyday common sense knowledge as well and more generally develop a reflexive posture which will make them less vulnerable to any kind of indoctrination including that of their science teachers.

References

American Association for the Advancement of Science. (1989). *Science for all Americans: Project 2061*. Washington, DC: AAAS.

Bauer, M. W., Petkova, K., & Boyadjieva, P. (2000). Public knowledge of and attitudes to science: Alternative measures that may end the "Science Wars." *Science, Technology, & Human Values, 25*, 30–51.

Collectif chimie Cégep Limoilou. (1998). *Pertageons nos connaissances: Des élèves rencontrent la population des quartiers populaires*. Québec: Cégep Limoilou.

Cross, R.T. (1997). Ideology and science teaching: Teachers' discourse. *International Journal of Science Education. 19*, 607–616.

Cross, R. T., & Price, R. F. (1992). *Teaching science for social responsibility*. Sydney: St. Louis Press.

Désautels, J. (1998). Une éducation aux sciences pour l'action. *Recherches en Soins Infirmiers, 52*, 4–11.

Eisenhart, M. A., & Finkel, E. (1998). *Women's science: Learning and succeeding from the margins*. Chicago: University of Chicago Press.

Eisenhart, M., Finkel, E., & Marion, S. F. (1996). Creating the conditions for scientific literacy: A re-examination. *American Educational Research Journal, 33*, 261–295.

Epstein, S. (1995). The construction of lay expertise: AIDS activism and the forging of credibility in the reform of clinical trials. *Science, Technology, & Human Values, 20,* 408–437.

Epstein, S. (1997). Activism, drug regulation, and the politics of therapeutic evaluation in the AIDS era: A case study of ddC and the 'Surrogate Markers' debate. *Social Studies of Science, 27,* 691–726.

Foucault, M. (1971). *L'ordre du discours.* Paris: Gallimard.

Fourez, G. (1997). Scientific and technological literacy as a social practice. *Social Studies of Science, 27,* 903–936.

Fuller, S. (1997). *Science.* Buckingham, England: Open University Press.

Guston, D. H. (1999). Evaluating the first U.S. consensus conference: The impact of the citizens' panel on telecommunication and the future of democracy. *Science, Technology, & Human Values, 24,* 451–482.

Jordan, B. (1989). Cosmopolitical obstetrics: Some insights from the training of traditional midwives. *Social Science in Medicine, 28,* 925–944.

Lave, J. (1977). Tailor-made experiments and evaluating the intellectual consequences of apprenticeship training. *Quarterly Newsletter of the Laboratory for Comparative Human Development, 1,* 1–3.

Lave, J. (1993). The practice of learning. In S. Chaiklin & J. Lave (Eds.), *Understanding practice: Perspectives on activity and context* (pp. 3–32). Cambridge, England: Cambridge University Press.

Lave, J., & Wenger, E. (1991). *Situated learning: Legitimate peripheral participation.* Cambridge, England: Cambridge University Press.

Lessard, N. (1989). *Une étude ethnographique d'un laboratoire de chimie en contexte scolaire: activités expérimentales ou rituelles?* Mémoire de maîtrise non publié, 265 pages. Université Laval.

McGinn, M. K., & Roth, W.-M. (1999). Towards a new science education: Implications of recent research in science and technology studies. *Educational Researcher, 28*(3), 14–24.

National Research Council (1996). *National science education standards.* Washington: National Academy Press.

Rabeharisoa, V., & Callon, M. (1999). *Le pouvoir des malades*. Paris: Écoles de Mines.

Roth, W.-M., & Alexander, T. (1997). The interaction of students' scientific and religious discourses: Two case studies. *International Journal of Science Education, 19*, 125–146.

Roth, W.-M., & McGinn, M. K. (1997). Deinstitutionalizing school science: Implications of a strong view of situated cognition. *Research in Science Education, 27*, 497–513.

Roth, W.-M., & McGinn, M. K. (1998). >unDELETE science education: /lives/work/voices. *Journal of Research in Science Teaching, 35*, 399–421.

Rowe, G., & Frewer, L. J. (2000). Public participation methods. *Science, Technology, & Human Values, 25*, 3–29.

2

Linking School Science Education with Action

Edgar W. Jenkins

Attempts to link the school science curriculum with social, political or economic concerns are as old as science education itself. Economic and utilitarian concerns were in evidence when science was first schooled in the mid-nineteenth century, although they were necessarily subordinated to the politically more resonant claims of liberal education. The general science movement of the years between the two world wars sought both to broaden the content of school science courses and to challenge the academic tradition upon which they rested. In seeking to "furnish the mind and give some knowledge of the world in which we live . . . without obtruding the notions of discipline and training in method" (Tilden, 1919, p. 12), the aim was not to transform that world but to enrich students' understanding of it.

A much more overt political agenda underlay the attempts of a small group of committed left-wing scientists in England during the 1930s to reform school science education in a direction that reflected their socialist beliefs about the role that science should play in society. Radical scientists like J. D. Bernal, J. B. S. Haldane, Joseph Needham and Lancelot Hogben derided the social and political connotations of pure science, arguing for the reorganization of science in the service of the state and for the contribution that science could make to citizenship understood in socialist terms. Thus, it should be less important for a biology teacher to learn about "ciliary current in Ammocoetes or about the crystalline style of Pelecypods than to know about vitamins and

nutritional diseases of babies or about iodine and calcium in relation to pig culture" (Hogben, 1942, p. 273). This attempt to accommodate curriculum change within a wider movement of political and social reform and the opposition to the movement itself
have been well documented (e.g., Werskey, 1978). Although the
movement collapsed after the end of World War II, its curriculum
legacy can still be felt. For example, there are several aspects of
the "social relations of science" (to borrow a contemporary term)
within courses bearing such titles as "social biology," "human biology," or "health education" (Jenkins, 1979).

The academic tradition of secondary education was strongly
reasserted after the war, finding its clearest expression in programs such as CBA and PSSC in the United States and the Nuffield O- and A-level science curriculum projects in the United
Kingdom. In these initiatives, students were to be encouraged to
"think like scientists" or to "behave like a scientist for a day."
Scientific knowledge was presented as the outcome of disinterested inquiry into the natural world and, in consequence, as objective, universal, free from considerations of value and uncon-
strained by context. In addition, science underpinned technology
and claimed hegemonic authority over it, although the conative
dimensions of technological activity were ignored.

Such a vision, although beguiling, was soon overtaken by profound shifts in the social and political context of science and of
science education. Rachel Carson's *Silent Spring*, published in
1962, prompted concern over the relationships between science,
technology and society, although conservation and other environmental issues can claim a much longer history. That concern has
since been fuelled by a succession of human and environmental
tragedies. These include issues arising from advances in such
fields as DNA and communications technology and from a di-
minishing confidence in the ability of science to provide in a variety of contexts the secure data or solutions it once seemed able to
promise. More profound are the commercialization and industrialization of the scientific endeavor and its integration with technology. This merger of knowledge with power has changed the relations between science, technology, production, and society at
large, and it has transformed the intellectual, instrumental and

organizational dimensions of scientific research. In short, science
has become worldly in every sense of the word.

Equally profound have been changes in scholarly understanding of the nature of the scientific endeavor. The work of historians,
philosophers, and sociologists of science has rendered "antique"
the understanding of science that "explicitly or implicitly [has]
provided coherence and security for generations of [science]
teachers" (Ravetz, 1989, p. 20). Such work has yet to have much
impact on the daily routines of science teachers, and it has not
gone unchallenged (Gross & Levitt, 1994; Sokal & Bricmont, 1998).
Nonetheless, it has undermined notions such as objectivity and
truth and, thereby, the authority traditionally vested in the scientific endeavor.

The principal curriculum response to the changed social and
political context of science has been the development of a variety
of courses commonly labeled "STS" (science, technology and society). Courses of this kind seem to hold the potential both to reas-
sert and to redefine the contribution that science might make to
functioning effectively as an individual and as an informed citizen
in a democracy. The reality has often been somewhat different.
For the most part, STS materials have lacked an adequate theoretical foundation and have served a limited, if valuable, function
of supporting and enriching otherwise conventional school science
courses. To this extent, they can be seen as attempts to rescue science curricula in difficulty, rather than as attempts to reshape
school science education in any fundamental way. They have not
impinged upon the technical and instrumental rationality and pre-
sumed certainty of science. Thus, "SATIS cannot improve public
perceptions of science overnight but . . . it can equip tomorrow's
scientists to argue their case more cogently than previous generations have succeeded in doing" (Nash, 1990, p. 1). At the start of
a new millennium, "improving the public perception" of science
sustains a large and thriving industry that includes interactive science centers, museums and exploratories, together with the print
and broadcast media. For the most part, however, science centers
and museums do not challenge or interrogate aspects of the scientific endeavor, and the few attempts that have been made to do
so have led to considerable political difficulties with funding

agencies or other interested parties (Macdonald, 1998; West, 1996).

Other curriculum responses include the development of courses in the "public understanding of science" (e.g., De Vos & Reiding, 1999; NEAB, 1998), attempts to teach scientific concepts through their everyday applications (Eijkelhof & Kortland, 1988), and individual or collective initiatives. These include curricula such as *Teaching Science for Social Responsibility* (Cross & Price, 1992) and *Chemistry in the Community* (ChemCom) developed by the American Chemical Society (ACS, 1998). *ChemCom* is a "year-long chemistry course designed for the bright student not planning to attend college" (OECD, 1993, p. 59). It seeks to emphasize the application rather than the generation of chemical knowledge and to highlight science in the community rather than on the laboratory bench. To these ends, decision making is privileged above mastery of subject matter, and the pedagogy reflects a commitment to student-centered, cooperative learning strategies such as role-play, surveys, interviews, and debates. Although these and other similar initiatives are important, they are commonly regarded as most appropriate for those who do not wish to pursue the study of science beyond school. Their status, therefore, is subordinate, and they leave unchallenged the hegemony of science programs that are more academic and "internalist" in approach and perspective.

If, in general, science courses constructed around such notions as relevance, teaching from context or via applications, social betterment and citizenship have effected, at best, only short-term or marginal changes in secondary school science education, it becomes important to ask why this is the case. No doubt, much can be explained in terms of the allegiance of many secondary school science teachers to their scientific disciplines, the prevailing power relationships within society and the unwillingness of the professional, and especially the academic, scientific community to address the implications of science for all. I want to suggest, however, that two other factors may be involved in the relative lack of success of these earlier attempts to link school science education with the world of practical action. First, they have accommodated a partial and under-theorized view of the role that science plays in the modern world. Secondly, such engagement has been vicarious rather than personal, and has paid insufficient attention to the wider consequences of linking science education with action. I discuss each of these factors in greater detail.

Science Comes in All Shapes and Sizes

Despite many attempts at reform, some more radical than others, much of secondary school science education continues to portray science in what might be called its heroic Enlightenment mode. School science is almost exclusively concerned with basic or fundamental science, driven by intellectual curiosity without thought of possible applications. Its heroes are represented by the likes of Newton, Faraday, Maxwell and Darwin, its philosophy is essentially positivist, and its history is essentially triumphalist in tone. Much might be said to defend this picture. Scientific imagination is all too easily undervalued, especially by those who appear to take some pride in neither understanding nor exercising it, and science has much to be triumphant about. In addition, rejecting positivist ideas in the absence of any alternative and well-confirmed picture of how science works does not come without danger. As Poincaré observed almost a century ago, "If the confidence that his methods are weapons with which he can fight his way to the truth were taken from the scientific explorer, the paralysis of those engaged in a hopeless task would fall upon him" (Heilbron, 1983, p. 178).

Today, however, basic or fundamental science is a relatively small part of the scientific research endeavor, and the notion that natural science stands pure and separate from all involvement with society has largely disappeared. To use the USA as an example, over half of the country's research scientists are located in industry and much scientific research "is social rather than theory driven" (Hurd, 1990, p. 413). For some commentators, aspects of this commercialization and industrialization of science constitute a new system of knowledge production (Gibbons *et al.*, 1994). That system is characterized by the emergence of *ad hoc* and more fluid institutional structures, the generation of knowledge in the context of use and the validation of such knowledge by criteria other than peer review. Another important aspect of scientific research in the modern world relates to the work of regulatory and legislative agencies, standard setting organizations and expert commissions, a field sometimes referred to as "mandated science" (e.g., Levy, 1989).

The distinctions between the three categories of fundamental, strategic, and mandated science are not absolute, but neither are they arbitrary. The scientific research associated with each category is concerned with different kinds of questions, different research methodologies and different criteria for validating the answers. Fundamental science is concerned with understanding natural phenomena and, where possible, establishing and explaining relationships. What processes are involved in leaf fall, in aging, or the development of a cancer cell in otherwise healthy tissue? Does the rate of a chemical reaction depend upon temperature? If so, what is the quantitative relationship between reaction rate and temperature? And is this relationship the same for all reactions or not? How can any differences or commonalities be understood in molecular or other terms? What is the qualitative and quantitative nature of the relationship between electricity and magnetism, force, and acceleration or gravity and light?

In contrast, strategic science attaches priority to the exploitability of scientific knowledge. The emphasis is on establishing a body of scientific understanding that can support "a generic (or enabling) area of technological knowledge; a body of knowledge out of which many specific products and processes may emerge in the future" (ACARD, 1986, p. 11). The research agenda is set by private industry and by governments, each of which necessarily makes decisions about how best to invest large, sometimes very large, sums of money in scientific research in the hope of eventual substantial commercial or other rewards. Countries and corporations differ in their mechanisms for foresight activity in strategic science, that is, in how they choose which areas of scientific research to prioritize for investment and, not surprisingly, in the choices that they eventually make. The scientific research eventually undertaken is not principally directed towards new and fundamental insights of the kind associated with science commonly described as pure or basic, although such insights may emerge. Research problems are often "field generated." They might include improving voice-recognition systems for computers or the information-carrying capacity of a so-called "smart material." They might also concern enhancing the local effectiveness of a drug, developing the capacity of systems to recognize objects, or improving the versatility and reliability of navigation systems. Or they might extend the range, sophistication, and effectiveness of tele-diagnosis and tele-medicine. Access to research and its outcomes is often protected in some way.

Mandated science is concerned with the construction of expert advice about a range of issues or the establishing of standards that can be enshrined in legislation or codes of practice and, where necessary, enforced by the courts. How are maximum, short-term, long-term, or occupational standards determined and defined for a range of hazardous materials, and why do these sometimes differ from one country to another? At what point can a food additive or drug be declared unsafe? When is noise or other environmental assault a threat to human health, and is the answer the same for children as for adults, for men as for women? Mandated science is thus often intimately related to the notion of risk assessment, an undertaking that has a long history in a variety of fields such as chemical engineering and systems design. Risk assessment in contexts in which science is involved, such as BSE or releasing genetically modified organisms into the environment, brings together the uncertainty and complexity of the real world in a way that school science, derived essentially from the first of the three categories identified above, seeks to avoid. Of its general importance, there can be little doubt. Thus, "the question of 'safety' is emerging as a great task for science, as a successor to the twin goals of knowledge and power that were articulated nearly four centuries ago in the Scientific Revolution" (Ravetz, 1990, p. 12).

Its more particular importance to school science education derives not only from this wider authority but also from the fact that it is in this context of uncertainty and complexity that most pupils encounter science-related issues outside school. What is the risk of contracting lung cancer from passive smoking? What is the threat to health of using the contraceptive pill, living near a nuclear power facility or taking a recreational drug such as cannabis? The last of these is a particularly interesting example since medical and legal opinion differs within and between countries or communities. It is with respect to questions such as these that many pupils are likely to see science as both useful and relevant to their everyday lives. The same context also allows them to see science as responsible for new, unanticipated and often global dangers, and lacking the prophylactic qualities that the Enlightenment attributed to it.

Science thus fulfils a variety of roles in the contemporary "risk society" (Beck, 1995). It manages to be seen simultaneously as villain and savior and as solution and problem. It is perhaps time that school science explicitly acknowledged these roles and accommodated their consequences for the form and content of the curriculum and for the way in which science is taught.

The Vicarious, the Personal, and the Practical

As noted above, attempts to make school science courses more relevant have a long ancestry and have taken many forms. In recent years, as the authority of science has increasingly been challenged and students have shown a marked reluctance to pursue advanced courses of study in the physical sciences, these attempts have acquired an added sense of urgency. None, however, seems to have taken much account of the research that has been done into how science is encountered and used in everyday life or into what is involved in "applying" scientific knowledge to the solution of problems. The relevant literatures are concerned, respectively, with the public understanding of science and with the nature of technological activity, and, more widely, of problem solving.

The relationship of lay citizens and other non-experts to science is much more complex than that normally captured by quantitative surveys of "public understanding of science" (Irwin, 1995; Irwin & Wynne, 1996). In particular, the relationship is interactive and interrogative and not one that can be conceptualized in the beguilingly simple terms of ignorance or rejection of scientific knowledge or expertise. There is now a well-attested body of empirical evidence that in the everyday world of action, science emerges not as coherent, objective, and unproblematic knowledge but as uncertain, contentious, and often unable to provide answers to the many important questions with the required degree of confidence. In some instances, and for legitimate reasons, scientific knowledge may be marginalized or ignored. In addition, such knowledge, assuming it exists, is not separated from its social or institutional connections, and it is weighed alongside other more personal or local knowledge in establishing a basis for action.

> "Citizen thinking," i.e. everyday thinking, turns out to be much more complex and less well understood than scientific thinking and, as might be expected, well adapted to decision-making in an everyday world which, unlike science itself, is marked by uncertainty, contingency and adaptation to a range of uncontrolled factors. (Jenkins, 1999, p. 704)

In the context of a scientifically literate citizenry, "citizen thinking" is intimately related to the notion of "citizen science," that is, science which relates in reflexive ways to the concerns, interests and activities of citizens as they go about their everyday business.

This perspective on the public understanding of science has a number of implications for attempts to link school science education with action. Most obviously, it is necessary to choose problems that engage the attention of students upon terms that they themselves find appealing and satisfying. Many attempts to make school science education more "relevant" have floundered on just this point (Newton, 1988). Students are unlikely to be interested in a scientific idea if they are not also interested in the application presented to illustrate it. Those who are bored by an everyday application of a potentiometer, lens, or thermostat are not likely to be attracted by the underlying physics, although there are always some students who will find the latter much more appealing than the former. Equally important is an acknowledgement that real-life problems are much more complicated than problems encountered in conventional school science courses. This is not simply a matter of choosing the "right" problem. School science courses are not directed towards action but to developing students' understanding of science itself. Any problems presented to the students, therefore, tend to be vicarious rather than personal, and the goal is primarily that of understanding the relevant scientific knowledge, not the solution of the problems. The constraints, resources, time-scale, and expertise available are also different, as are the criteria for determining the success of the outcome. In addition, real life problems are rarely well bounded, and their solution requires knowledge that extends beyond the physics, chemistry, and biology of school science courses.

To this extent, there is much in common between "citizen thinking" and the public understanding of science, on the one hand, and the technological knowledge that sustains practical capability, on the other. Technological knowledge is not the same in form and substance as scientific knowledge, and it is structured by the tension between the demands of functional design (i.e., it enables some purpose to be achieved) and the particular constraints of the context of working. Those who seek to draw upon scientific knowledge for practical purposes tend to do so quite pragmatically and indiscriminately, without regard for the coherence and progression that characterize traditional science courses. They sometimes draw upon concepts and models, which, while adequate for the purpose in hand, command no fundamental scientific validity. For example, workers in a computer company, chained to their benches by a grounded metal bracket in order to prevent damage to sensitive electrical components by the build-up of static electricity have particular views about electricity. They regard electricity as a fluid, which can either pile up (static electricity) or be discharged to ground, where it is dispersed or "lost" (Caillot & Nguyen-Xuan, 1995). This "unscientific" model of electricity nevertheless enables the workers to function safely and to make sensible decisions when confronted with problems. Similarly, heating engineers invoke a fluid theory of heat, reminiscent of the 18th-century caloric model, when discussing a range of problems related to heat transfer and conservation.

Those engaged in practical problem solving also invent concepts that do necessarily have scientific counterparts or are open to criticism on the grounds that they are empirical, rather than fundamental. Layton (1993) provided a number of examples, such as interior lighting designers' use of concepts such as "discomfort glare" and "disability glare." These notions do not feature in school textbooks on light. He also makes the point that

> Those who wrestled . . . with the design of steam engines would have found their most valuable theoretical guidance in the Count de Pambours' (1837) *Théorie de la machine à vapeur* [whereas] the theory of his contemporary Sadi Carnot was independent of design features, working substances and mechanisms . . . and did not relate to the practicalities of building an engine which could be relied upon. (Layton, 1993, p. 145–146)

Given the above, it is hardly surprising that most teachers are likely to lack the "knowledge in action" needed to help students connect everyday experiences with what goes on in their science lessons (Cajas, 1999). Schools do not emphasize knowledge in the context of use, and school science stresses the universality and decontextualized nature of scientific understanding. Changing this situation offers a major challenge to the "Euro-American institution of schooling which promoted an individually centered analytical approach to tools of thought and stresses reasoning and learning with information considered on its own ground, extracted from practical use" (Rogoff, 1990, p. 191). Beyond this, the integration of knowledge with action suggests that the development of students' practical understanding has less to do with the mastery of universal scientific principles and concepts than with apprenticeship. That is, it has less to do with induction into the processes whereby novices become members of a community of practice, each of which is engaged in its own replication and reproduction (Lave & Wenger, 1991).

An indication of what Rogoff's challenge might mean for school science is evident from a number of initiatives that have been taken to develop what might be called practical reasoning, that is, acting wisely in practical matters, in environmental education. This presupposes that environmental education involves more than the acquisition of knowledge about a range of environmental matters and is concerned with the generation and implementation of solutions to practical problems or needs. Given this, environmental education exposes with particular clarity the complex interactions among social, economic, personal, and other value positions associated with almost any environmental issue such as acid rain, ozone depletion, or atmospheric pollution. Interestingly, research offers little support for the idea that the ways in which people treat the environment can be modified by helping them to acquire the relevant environmental knowledge or attitudes (Hungerford & Volk, 1990). What seems to be lacking is the engagement of such knowledge and attitudes with the commitment to action in relation to the environmental issue concerned. Commonly, such an issue is initially of local, rather than national, international or global concern, although the local issue may be a particular example of a wider problem. Sutti (1991) has reported a collaborative study (involving five local secondary schools in Italy) of the degree of pollution of surface and ground water in the communities within which the schools were located. The students were involved in chemical and microbiological analyses of water samples and in discussing their data with the water authorities. The outcome of their work was local knowledge, generated for a specific purpose within a particular context, directed towards a particular audience, and, ultimately, concerned with action. The importance of context for action is illustrated by another example of environmental education, this time from Sweden. Here students investigated two local lakes affected by acid rain. Working in collaboration with the local authorities, the students developed a program that generated a commitment among the local population to improving the condition of the lakes by adding lime. This commitment was important since it not only involved support for the students' initiative but also generated a confidence, previously lacking, among the local population that matters were capable of improvement. However, when other schools copied this idea, the program was reduced to the technical dimensions of liming frozen lakes. The imaginative and dynamic features of the original project, central to its educational function, were lost (Posch, 1993).

What general conclusions about linking science education with action might be drawn from these and other examples? School science will need to respond to the fact that establishing a basis for wise action requires a wider knowledge base than that provided by the conventional school disciplines of physics, chemistry, or biology. For such action, boundaries do not exist between the natural sciences or between natural science, social science, and other forms of knowledge and understanding. In addition, much of the science related to practical action is less than secure, even controversial. The science relating to any long-term physiological or other effects of cannabis—its possession or sale is illegal in most but not all countries—is an interesting example. It is not difficult to think of other examples such as the decision whether to permit or to ban a food additive, and the determination of a legal limit of blood alcohol below which it is "safe" to drive a motor vehicle. It follows that school science curricula in different countries are likely to show a greater degree of variety than is presently the case. Not all science-related issues are global, and if a school science curriculum is to be sensitive to the interests of students, regional or other in-country variations will need to be accommodated. It may also be necessary to recognize different interests on the part of boys and girls and to respond appropriately.

Such an approach to the curriculum has obvious pedagogical implications. Traditionally, school science is well bounded and presents science as well established and secure. Contested science, matters such as risk and social or legislative policy, has been excluded. If these are to be accommodated, time will need to be made available, and some of school science teaching will need to be recast. Some traditional and familiar practices will need to be abandoned in favor of strategies which involve engagement with issues which are controversial, messy and have to be brought into focus only to lack a unique or even (initially or eventually) agreed solution. These, of course, are the characteristics of much reasoned practical activity. More particularly, science teachers may need to develop some of the skills more commonly associated with other subjects of the school curriculum, e.g., history or English, in which debate, controversy, and alternative outcomes constitute more familiar territory.

There are also implications for legislators and policy makers within science education, since linking science education with action requires a degree of flexibility in, and local control over, at least some of the science curriculum. Unless carefully framed, centrally prescribed science curricula, often buttressed by statutory provisions of assessment to raise standards or enhance accountability, are unlikely to allow schools and science teachers the freedom they need to work in this way. Already, in England and Wales, there is evidence that the national curriculum is preventing many science teachers from teaching in ways that they judge to be in the best interests of their students (Donnelly & Jenkins, 1999).

Linking science education with action also presents severe problems for assessment. Posch (1993), for example, has quoted a group of Austrian students who regarded grading by their school of their environmental work as a "devaluation" of what they had done. Their assessment criterion was the "real-life" evaluation they had encountered in dealing with the people with whom they came into contact during their environmental project in the community. In a similar way, the grade-7 students in Roth and Lee's study (this volume) found the real evaluation of their work in the successful exhibit of the environmental activists' open house event. The response of these students has an important message for those who seek to assess knowledge in action by reference to an abstracted, even reified, procedural account of what is supposed to be involved in acting wisely in the realm of the practical (Atkin & Helms, 1993). It also serves as a reminder of the more general importance of not underestimating student reaction to science curriculum change (Delamont, 1990).

In seeking to weaken the traditional insulation of the school from the "real" world beyond, science education for action helps to fuel a broader debate about the nature and purpose of schooling in a world increasingly shaped by information technology and confronted by environmental and other problems of global significance. It also privileges local understanding at the expense of the universal and, thereby, threatens to discredit the latter because of its association with particular power structures (Foucault, 1970). Additionally, by highlighting ambivalence towards scientific knowledge as a universal "given," scientific understanding in the context of action may claim a post-modernist flavor. Those who wish to use scientific knowledge to ground their actions inevitably find themselves, like many postmodernists, questioning much of what was presented as "given." They discover that, in some cases, scientific 'facts' involve a large element of subjectivity or relate to circumstances far removed from those in which they seek to act.

From another perspective, the relationship between scientific knowledge and other forms of local and particular knowledge amounts to a challenge to an epistemology of science that insists that the world is separate from the scientific observer. An alternative to the "objectivity" that this epistemology presumes is to adopt a phenomenological stance and accept a world that is both objective and participatory. This allows "truth within situations" (Prigogine & Stengers, 1984). The selecting, evaluating, and constituting of knowledge, including scientific knowledge, for the purposes of action can be described in this way, although the term presents more problems than "situated cognition" (see Roth & Lee, this volume) with which it obviously has much in common.

Ultimately, linking science education with action demands a rationale which is fundamentally different from that which underpins the science curricula traditionally offered by schools and colleges, and it is essentially about the empowerment of students as citizens. To this extent, it is also about political education, a goal about which schools have historically been very cautious and one more likely to command hostility than support from the professional scientific community. The challenge therefore is profound. The question of whether science education for action is about knowledge or activism is but a particular version of the fundamental question "What is school science for?"

References

Advisory Council for Applied Research and Development (ACARD). (1986). *Exploitable areas of science*. London: HMSO.

American Chemical Society (ACS). (1998). *Chemistry in the community*. Dubuque, IA: Kendall/Hunt.

Atkin, J. M., & Helms, J. (1993). Getting serious about priorities in science education. *Studies in Science Education, 21*, 1–20.

Beck, U. (1995). *Ecological politics in an age of risk*. Cambridge, England: Polity Press.

Caillot, M., & Nguyen-Xuan, A. (1995). Adults' understanding of electricity. *Public Understanding of Science, 4*, 131–152.

Cajas, F. (1999). Public understanding of science: Using technology to enhance science in everyday life. *International Journal of Science Education, 21,* 765–774.

Carson, R. (1962). *Silent spring.* London: Hamilton.

Cross, R. T., & Price, R. F. (1992). *Teaching science for social responsibility.* Sydney: St. Louis Press.

De Vos, W., & Reiding, J. (1999). Public understanding of science as a separate subject in secondary schools in the Netherlands. *International Journal of Science Education, 21,* 711–720.

Delamont, S. (1990). A paradigm shift in research on science education? *Studies in Science Education, 18,* 153–158.

Donnelly, J. F., & Jenkins, E. W. (1999). *Science teaching in secondary schools under the National Curriculum.* Leeds, England: Centre for Studies in Science and Mathematics Education/Centre for Policy Studies in Education, University of Leeds.

Eijkelhof, H. M. C., & Kortland, K. (1988). Broadening the aims of physics education. In P. J. Fensham (Ed.), *Development and dilemmas in science education* (pp. 282–305). London: Falmer.

Foucault, M. (1970). *The order of things: An archaeology of the human sciences.* New York: Pantheon.

Gibbons, M., Limoges, C., Nowotny, H., Schwartzman, S., Scott, P., & Trow, M. (1994). *The new production of knowledge. The dynamics of science and research in contemporary societies.* London: Sage.

Gross, P. R., & Levitt, N. (1994). *Higher superstition: The academic left and its quarrels with science.* Baltimore: Johns Hopkins Press.

Heilbron, J. L. (1983). The virtual oscillator as a guide to physics students lost in Plato's cave. In F. Bevilacqua & P. J. Kennedy (Eds.), *Proceedings of the International Conference on Using History of Physics in Innovatory Physics Education* (pp. 162–182). Pavia: University of Pavia.

Hogben, L. (1942). Biological instruction and training for citizenship. *School Science Review, XXIII,* 263–281.

Hungerford, H. R., & Volk, T. L. (1990). Changing learning behaviour through environmental education. *Journal of Environmental Education, 21,* 8–12.

Hurd, P. D. (1990). Guest editorial: Change and challenge in science education. *Journal of Research in Science Teaching, 27,* 413–414.

Irwin, A. (1995). *Citizen science: A study of people, expertise and sustainable development.* London: Routledge.

Irwin, A., & Wynne, B. (Eds.). (1996). *Misunderstanding science? The public reconstruction of science and technology.* Cambridge, England: Cambridge University Press.

Jenkins, E. W. (1979). *From Armstrong to Nuffield: Studies in twentieth-century science education in England and Wales.* London: Murray.

Jenkins, E. W. (1999). School science, citizenship and the public understanding of science. *International Journal of Science Education, 21,* 703–710.

Lave, J., & Wenger, E. (1991). *Situated learning: Legitimate peripheral participation.* Cambridge, England: Cambridge University Press.

Layton, D. (1993) Science education and praxis: The relationship of school science to practical action. In E. W. Jenkins (Ed.), *School science and technology: Some issues and perspectives* (pp. 118–159). Leeds: Centre for Studies in Science and Mathematics Education, University of Leeds.

Levy, E. (1989). Judgement and policy: the two-step in mandated science and technology. In P. T. Durbin (Ed.), *Philosophy of technology: practical, historical and other dimensions* (pp. 41–59). Dordrecht, Netherlands: Kluwer Academic Press.

Macdonald, S. (Ed.). (1998). *The politics of display.* London: Routledge.

Nash, I. (1990). No more mad scientists. *Times Educational Supplement,* 2 November, p. 10.

Newton, D. (1988). Relevance and science education. *Educational Philosophy and Theory, 2,* 7–12.

Northern Examinations and Assessment Board (NEAB). (1998). *Science for public understanding.* Manchester, England: Author.

Organization for Economic Cooperation and Development (OECD). (1993). *Science and mathematics education in the United States: Eight innovations.* Paris: Author.

Posch, P. (1993). Research issues in environmental education. *Studies in Science Education, 21*, 21–48.

Prigogine, I., & Stengers, I. (1984). *Order out of chaos: Man's dialogue with nature*. London: Heinemann.

Ravetz, J. R. (1989). Some new ideas about science relevant to education. In E. W. Jenkins (Ed.), *Policy issues and school science education* (pp. 41–59). Leeds: Centre for Studies in Science and Mathematics Education, University of Leeds.

Ravetz, J. R. (1990). *Beyond the good and the true in science education*. Unpublished paper.

Rogoff, B. (1990). *Apprenticeship in thinking: Cognitive development in social context*. New York: Oxford University Press.

Sokal, A., & Bricmont, J. (1998). *Fashionable nonsense*. New York: Picador.

Sutti, A. (1991). The water analysis project: an alternative model for environmental study. In OECD/CERI, *Environment, schools and active learning* (pp. 59–65). Paris: OECD.

Tilden, W. (1919). *School Science Review, 1*, 12.

Werskey, G. (1978). *The visible college*. London: Allen Lane.

West, R. M. (1996). ACS abandons negotiations with Smithsonian over science in American Life exhibition. *The Informal Science Review, 17*, 9.

CLUSTER I

Science Education in/through Environmental Activism

Conversation 1

Jacques: You know, Jenkins is like a mapmaker who charts the territory that we are broaching in this book. The topics he raises are taken up in the different chapters with increasing depth, even though they do not necessarily follow the same orientation or development that he would have proposed himself. But he has sufficient experience to know that this long discussion on quite problematic subjects will not be fruitful if the participants cannot express divergent viewpoints. Then, the consultation of our elder did not consist in asking about the fine details of the best path to follow but rather to inquire about some of the dead ends he has encountered and some of the promising avenues he foresees.

Michael: Your suggestion of using conversations to talk the reader from section to section in this book makes me think of the process followed by the music composer Modest Mussorgsky to organize "Pictures at an Exhibition." In this set of pieces, he not only gives us his impressions of each picture but also walks us from one picture to the next. If we follow your idea of Jenkins as a mapmaker charting the landscape of issues and promising avenues, you and I are the guides taking the reader from one major region to the next.

Jacques: I agree with this idea. In fact it gives us the room to maneuver around certain epistemological obstacles. For instance a good number of books are structured along the traditional distinction between educational theory and practice. So in designing the landscape according to a set of issues we are implicitly contesting the value of this distinction. Indeed, I think that, we should rather think of knowing in terms of practice of which the epistemological practices in the third chapter cluster in this book are but one type of practice among many.

Michael: The two chapters in the present cluster are a good illustration of what you are saying. One of them illustrates a more empirically oriented practice and the other one a more theoretically oriented practice. But there are other reasons for regrouping these two chapters in the first cluster. Somehow they illustrate that enacting science education as a form of social action is not a utopia. The detailed account of the different social actors' activities and of the transformative processes within the community is a vivid and convincing illustration of the general thesis of this book.

Jacques: I think we made a good move in inviting the reader to first take notice that we are involved in empirical practices and not merely discuss, in our university offices, what ought to be done. This is all the more the case, since the chapters also show that you cannot detach scientific knowledge from other human practices including politics, aesthetics, and ethics.

Michael: It is true that over the past 30 years, and especially with the rise of post-modern and feminist thought, the traditional character of science—neutral, without passion, quasi-transcendental and, some would add macho, has been brought increasingly into relief. In this first set of chapters, we see how science and science education are enacted as another set of practices. What is crucial to me, then, are the values and ethical commitments within which all science-related activities are couched …

Jacques: … but are never taught in the traditional curriculum concerned with cold facts and theories.

3

Learning Science in the Community

Stuart Lee & Wolff-Michael Roth

We live in a time when tremendous technological extension of human abilities and lifeworlds[1] is having a profound effect on our culture, neighborhoods and environment. Even with all this power available to them, many people remain on the outside of the technological riptide surging through society. Some science educators claim that this is evidence of a severe problem of technological and scientific illiteracy (Hazen & Trefil, 1991). Unless our students become scientifically literate, these educators say, they will have great obstacles to becoming empowered participatory citizens integrated meaningfully into society.

Although we may agree that there is a societal problem of scientific and technological illiteracy, we take a different view of both the problem and potential solutions. For us, common examples used to "demonstrate" illiteracy fail to be convincing. For example, the film *A Private Universe* showed Harvard Ph.D. graduates who explained summer heat and winter cold in terms of the sun-earth distance rather than the inclination of the sun's rays. In response to claims that this shows illiteracy, we reply that the inference is quite reasonable (it is valid for asteroids), and the fact

---

1. The term "lifeworlds" refers to the entirety of lived experience, as a being in the world and draws its conceptual basis from Heidegger's phenomenology (Heidegger, 1977). For example, television news extends our lifeworld far beyond our everyday experience.

that this interpretation is wrong in the context of the earth has little relevance to the graduates' lives. Another standard example is the difficulty many people experience in appropriately programming a VCR. In this case the problem could be considered one of poor design or technical writing (e.g., Winograd, 1996).

We are more interested in how people do "science" than whether they can recite fragments of scientific discourse or interpret engineers' technical inscriptions. When planning a meal, building a compost pile, taking a child for a nature walk, how does a person make sense of and act appropriately toward their physical surroundings?

We offer the thoughts of the Belgian philosopher, Gerard Fourez, as a different approach to scientific literacy. In an article on scientific and technological literacy, Fourez (1997) outlines some of the major practices one may engage in through the everyday use of science, whether as an environmental activist or an engineer. These frame his notion of scientific literacy, and include the skillful use of "experts; black boxes; simple interdisciplinary models (rationality islands); metaphors, comparisons and images; translations; standardized and disciplinary knowledge; and rationality in the process of making decisions." Using these criteria for assessing literacy, we suggest that science is very different and much more than what is currently taught in schools.

In this chapter, we explore the questions, What does "scientific literacy" look like? How does the experience of participating in an authentic science-related activity differ from that of learning science in school? and What is science? We scaffold our analysis of a case study on Fourez's (1997) notion of scientific literacy and Bourdieu's (1990) "logic of practice."

In the rest of this introduction, we critique current educational practices, focussing on two problematic aspects of science education. To suggest what we think scientific literacy looks like, we sketch anthropologists' findings of what technoscientific professionals do at work and touch on some of the common features of science in everyday life outside the work world.

School, Professional and Everyday Science

Is School Science Scientifically Literate?

Over the past decade, science education as enacted in schools has come under criticism because of the constraints to meaningful learning it constructs (Tobin, 1990; Tobin & Gallagher, 1987). We do not want to pursue this line of argument, although it is clear that little has changed since Tobin and his associates' first qualitative research projects (Roth, Boutonné, McRobbie, & Lucas, 1999). Here, we want to briefly address two issues seldom discussed but significant in the context of our recent work. First, science and other subjects in schools are clearly demarcated into different domains spatially, in separate rooms, temporally, in time slots and cognitively into distinct subjects. Second, science encourages the production of a submissive populace striving to please an authoritarian figure and discourages engaging in and challenging the current social political issues of the time.

Demarcations. In schools, the main narratives of science are divided into units of subject area and, within subjects, into groups of decontextualized "facts." In everyday scientific endeavors these divisions are not so clear. Relations become more important as people gather together the appropriate tools (advice from an expert in another field or a new detection method), irrespective of discipline, to solve a problem. Facts become less important than the contexts, which argue for or against the acceptance of knowledge claims. What controls were run? What was the sample size? How was the variable measured? Who did the work? These are the issues practitioners consider when working with "facts."

Schools also create boundaries between science and other aspects of society, such as economics, politics and history. Science is taught as a sterile, disinterested pursuit of "perfect" knowledge. It is not uncommon to hear science teachers tell their students "That's not science, that's politics. Let's keep out of politics" (see also the chapter by Edgar Jenkins, this volume). This is not consistent with sociological studies of science and technology, which have repeatedly shown science to be interwoven in a constitutive way with other aspects of life (Latour, 1993a, 1993b).

The dissociation of school science from other school subjects is also embodied in the physical structures of school life. Beginning with middle school, students often move into specialized rooms ("laboratories") where they enact particular practices said to develop their scientific skills. This physical separation demarcates science and associated practices. Laboratories become special places ("cathedrals" [Knorr-Cetina, 1992]), with their special almost religious routines, their own forms of discourse, forms of adulation, and so on (Fuller, 1997). Although some everyday professional science is performed in laboratories, much everyday science is practiced outside their bounds—in offices, kitchens, and backyards.

The temporal sequencing of the school day which jerks children out of an activity before they can develop a coherent sense of it or a conceptual context before they understand it both contribute to an experience of alienation from the subject.[2] These short time slots are also experiences of time that are not consistent with those experienced while practicing science. Science tends to progress slowly. It is prolonged (it commonly takes researchers days or weeks to answer a simple "yes or no" type of question and up to years to address more complex issues) and cyclical. The same procedures are enacted over and over again, the same sites visited, or the same samples analyzed.

With so many boundaries separating science as taught from both other subjects and the practice of science (formally and informally) in the everyday world, it is little surprise that science is often not experienced as a subject relevant to students' lives.

Creating a Compliant Populace Many aspects of school science devalue the lifeworlds the students bring to it. School science is taught as if it trained future working scientists (Roth & McGinn, 1997). This focus is reflected in the systematic presentation of the

---

2. In several innovative curricula, we could show how children build deep understandings in a variety of domains. These include simple machines (Roth, McGinn, Woszczyna, & Boutonné, 1999), ecology (Roth & Bowen, 1995), or structural engineering (e.g., Roth, 1998). When students are empowered to explore phenomena over long periods of time, driven by their own interests and intents, students tend to show tremendous intellectual competency.

detailed mechanics of theoretical constructs ("concepts," "scientific process"). These are comprehensible and of interest to only a small number of students (as Lemke [1990] showed, mostly white middle class) and are the "dress rehearsals" of routines of the lab where one is led through a standardized set of procedures. Little heed is paid to the fact that the very discursive forms and activity structures themselves are biased along the lines of traditional markers of difference such as gender, race and socio-economic status (Haraway, 1995; Roth & McGinn, 1998). This focus on scientists' science unrelated to the children's world effectively silences the students and encourages them to become passive and memorize meaningless (i.e., disconnected from what they know) bits of text.

Given that success and attendant rewards (marks, self-confidence, career options) depend on the selection (grading) mechanism, it is not surprising that the motivation driving students' learning of scientific minutiae is often to "get things right," to please the teacher. In science classes, people are judged more certainly to be right or wrong than in other subjects, such as English or philosophy where critical reasoning and original thought are given credit. This authoritarian emphasis on "getting it right" typically results in average students agonizing over whether or not they can guess the expectations of the instructor on exams and assignments, rather than being driven to explore an issue more deeply because of their passionate interest in it.

Even the practice of science through student labs becomes more of an effort by the student to "see" what they are expected to see and to do things the way they are supposed to be done (Roth et al., 1997) than an activity driven by their needs and curiosity. It becomes a ritual dress rehearsal rather than authentic practice (Bourdieu, 1990).

Finally, the system of rewards (marks) leads to a stratification, some say a rite of passage (Brookhart Costa, 1993) to select appropriate individuals, which ultimately benefits only the "successful" (compliant) students and those institutions which train and employ scientists: industry, government and universities. This symbolic violence excludes many students with interest and aptitude not measured by the regimented assessment system. It seems that both the subject matter and the method of instruction are not geared toward generating a scientifically literate populace, but rather function like a Fordian production line in a Foucauldian (disciplining) institution that forms employees of a certain class for a limited number of powerful institutions.

Laboratory Science, Science at Work

Over more than two decades, increasing numbers of ethnographic studies in scientific laboratories (and other similar environments) have shown science in a heretofore-unknown perspective. Science as practiced is very different from the way it is presented in schools and through the traditional, objectivist history of science (e.g., Fuller, 1997; Knorr-Cetina, 1981; Latour & Woolgar, 1986; Law, 1994; Pickering, 1995). Accordingly, science as practice is very context dependent in both a local sense (tools, people, discourses, local practices) and in a global sense (politics, economics, funding cycles, societal problems). Many "facts," to scientists, are understood to be dependent on many variables and reproducible under only the most controlled conditions. Rather than being self-evident, facts are often woven of strands selected carefully from complex webs of information.

The practice and experience of research is an activity well characterized by sociologists of science (e.g., Latour & Woolgar, 1986; Pickering, 1995). It is an active pursuit, typified by pursuing a question(s) of importance to the researcher. This practice involves framing questions, engaging with technology in a purposeful way to answer (or engage) these questions, and creating and circulating artifacts. It also involves engaging in discourse to solidify the interpretation of the artifacts, creating inscriptions around the artifacts, presenting the intellectual nexus at conferences, and convincing others of both the importance and validity of knowledge claims. The construction of facts is seen as a complex phenomenon, where whom you convince with your data is just as important as producing "conclusive evidence." In daily research practice, there is a high tolerance for and indeed an expectation that practitioners will make errors in the course of their work. Improvation, both due to technical constraints and shifts in understanding the problem, is the norm. Scientists are professionals, who need to be able to argue persuasively against opposition, act and think autonomously and take risks in their daily practice.

Science is also practiced at countless sites outside of laboratories. The fields of medicine, engineering, surveying, urban planning, law and marketing (among others) demand practices that use scientific techniques to examine the world at large and to justify appropriate actions. The boundaries between not only scientific disciplines but science and non-science blur at these sites. Science becomes one of a range of (discursive, material) resources available to practitioners in order to solve a problem.

Science in Everyday Life

In our culture, science is given primacy to define the "essential world" (Haraway, 1995). Unlike culture and politics, which are acknowledged to be the result of negotiation and to be constantly in flux, science promises us knowledge of a "real" world that stands beyond time and context. This timeless utility gives science a high position in our culture; it is called upon to provide instructions of how to do things and guidance on whether to do things. Even its linguistic register has become a powerful legitimizing rhetorical resource. Because of this, a great deal of political activity swirls around science—in many public debates the conflicting parties jostle for the right to claim science as their ally (e.g., Gieryn, 1996).

But the use of science in everyday life goes beyond the public debates. We all interact with science every time we wonder about the tests our doctor orders for us, or whether or not we should buy a water filter, or whether we take our babies out in the sunshine and for how long. It is a tool we use to make sense of and guide us in our decisions about the world.

But the flow of information is not just in one direction, from science to the masses. Science is also formed and informed by "real world" factors. Its metaphors are those of the culture in which it is embedded (the computational metaphor of neuropsychology, the "blueprint" metaphor of the genetic code). The problems it tackles are determined largely by societal expectations (the Manhattan Project, AIDS research) or those of business (genetically modified plant seeds, CFCs). And the kind of and pace of information it produces is determined in part by public or private funding agencies whose largesse gives scientists access to the equipment they use (Epstein, 1995, 1997).

Activists straddle a unique place in the science world. They enact science, they depend on it to define their goals and justify their claims, they influence its direction through politics and the media, and they work to involve citizens in its process and debates. The practice of activism can be thought of as the confluence of many of the different streams of science-related activity

In the remainder of this chapter, we provide a case study from our ethnographic research which tracks a suburban middle class environmental activist group as they work toward improving the health of a local stream and the watershed that feeds it. We present the work in the group as an authentic everyday science-based activity. By examining the learning that goes on within the group, we can tease out many aspects of learning that prove true to both the descriptions of scientists' scientific practice and science as it is enacted in everyday life. Here we confront people grappling with the issue of enacted scientific literacy as they use it rhetorically, to guide their physical actions, and to frame their understanding of the place in which they live.

Case Study: The Riffle in Gordon Creek

Our case study involves activities surrounding a watershed in the community where one of us (Michael) lives and in an on-going manner teaches in a local middle school (see Roth & Lee, this book), and where the other (Stuart) is a member in an environmental activist group. Our research is therefore enmeshed with community life. In this community, we have therefore been enacting participant-observer and observer-participant research, in particular with the community-based activist group, the Henderson Creek Watershed Restoration Project ("the Project"). As part of living and doing research in this community, we have come to re-alize that not only are our lives enmeshed with research, but that in everyday pursuits of people, science is irreducibly enmeshed with politics, aesthetics, farming, activism, and so forth.

As part of our work, we keep extensive fieldnotes, videotape Henderson Creek watershed-related activities, keep newspaper clippings related to the watershed, and interview local residents about their views on the activities surrounding the watershed. To provide readers with an idea of how science is enacted as part of living in a community, we report on the building of a riffle in one of the local parks traversed by the creek.

Background for Building the Riffle

The goal of the group is to "protect and enhance the Henderson Creek stream system" (Proposal, p. 23) to provide sufficient wa-ter for both the ecological and human needs of the watershed. As part of enhancing the stream system in the Henderson Creek wa-tershed, the group decided to introduce a series of riffles in a large tributary of Henderson Creek, Gordon Creek. The following is a presentation and analysis of some of the aspects of building the first riffle in Gordon Creek.

Project Proposal: An Undertaking Based in Science

Prior to building the riffle, the Project performed a surveying and sampling regime to map out the area they were interested in en-hancing. The results of this research were presented in a docu-ment, which was part of a package submitted to the municipality in order to secure permission to access Gordon Creek as it flows through Community Park.

This document includes color photos of the segments of the creek under contention, road maps, technical maps, aerial photo-graphs, and tables of stream characteristics (i.e., dissolved O_2 levels at different points along the stream). The stream is de-scribed in terms of features such as length, slope, mean-bankfull width, depth and width/depth ratio. Appendices listing the technical advisory committee—affiliates with the Pacific Geoscience Centre, Ministry of Environment, Lands and Parks, and the local university are included. The project has a strong association with Federal Marine Research Institute (FMR).

As they discuss different sections of the stream ("Reach 1, 2 and 3"), the linguistic resources drawn on are of a technoscientific nature.

> The lower 55 meters of this reach are relatively undisturbed, with vegetated banks, cobble- and gravel-dominated substrate, stable undercut banks, and 2 small pools.... The transition between Reach 2 and Reach 3 is marked by a decrease in slope from 2.6% to 1.7%, and increase in channel width, and a shift in bed composition from boulder, cobble and bedrock to sand, gravel, silt and cobble. The stream morphology becomes fairly straight and is composed of a 70 m long pool-riffle sequence downstream of the 16 m pool. (Proposal, p. 17)

Here, the activists clearly use a scientific repertoire (genre, register) to make an argument for building the riffle. There are not just numbers and metric units (though Canada has converted to the metric system, much of everyday life is dominated by the imperial system), but some of these numbers include decimal notation. The creek does not just have a width, but a "channel width"; the creek is not just filled with rocks, but there is a "bed composition," including boulders, cobble, bedrock, sand, gravel, and cobble. Thus, in writing their proposal to the government agency and to the community leaders, the activists draw on a scientific repertoire. Whether or not they had "prior knowledge" of such repertoires, they knew that their proposal would be more convincing if it drew on the appropriate repertoire, and in the appropriate form. As part of their riffle-related activities, they enacted science both in the content (register) and argumentative form.

Why a Riffle? Why the Park?

Thus goes the riffle discourse: Riffles are stream sections characterized by steeper slopes, higher velocity, and shallower water depth. In stream restoration work they are structures made of stone which are essentially artificial rapids. Their major purpose is to dissipate the energy of the stream's current, thereby reducing its potential to erode banks. They also serve to oxygenate the water; this increase is known to be good for two indicators of stream health, trout and benthic invertebrates. Thus, a riffle addresses a number of the key assessed issues[3] facing Henderson Creek—it helps to reduce siltation both through acting as a silt trap and by reducing the energy of the stream's current, which will reduce erosion of downstream banks, and it helps to oxygenate the water.[4] They create pools upstream of the riffles that can be layered with spawning surface—sand and gravel, and thus provide crucial spawning habitat for the fish. With the goal of stream restoration in mind, putting riffles in Henderson Creek makes good scientific sense.

But where along the creek should a riffle be inserted? The Project's steering committee chose the tributary of Henderson Creek, Gordon Creek, as it flowed through Community Park. Community Park is a local park of approximately 40 hectares, about half of which are wooded second growth stands of cedar, fir, hemlock, aspen and maple. As it runs through the park, the creek is protected by this over-story of vegetation, and for the most part, has sufficient bank-side vegetation to have stable banks. Although "adult cutthroat trout have been noted through the park... trapping through the park has failed to note a significant presence" (Proposal, p. 9). The park is directly upstream of the best cutthroat trout habitat in the entire stream system. This system is a stretch of about 1512 m of "natural, sinuous channel characterized by riffle-pool sequences, deep pools, stable undercut banks, overhanging vegetation, gravel beds, and habitat features such as large woody debris and boulders." "Trapping and electroshocking through these reaches indicate cutthroat through all age classes"

---

3. Henderson Creek is beset with a variety of problems common to streams passing through twentieth-century humanized landscapes. There is high siltation due to erosion, flash flooding and drying out due to storm sewers, fecal contamination from horses and cattle, channelization, denuding of the banks, re-routing, burying, pollution from storm sewers and industry, etc.
4. Trout need relatively high levels of dissolved oxygen to survive.

(Proposal, p. 7). If stream restoration could be performed successfully on the 600 meters of stream that runs through the park, it
would represent "a 34% increase in trout-rearing, spawning and
feeding habitat" (Proposal, p. 7). The fact that Gordon Creek in
Community Park, with many of the conditions necessary for trout
life, is contiguous with the best trout rearing habitat in the creek,
makes it an ideal location to begin restoration work.

As presented, this is a tightly woven scientific argument. In
fact, trout and the need for a riffle mutually stabilize each other
within the argument: Without trout, there is no need for a riffle,
but the riffle creates and improves habitat for the trout.

Networks and Actors

But scientific validity was not the only aspect of the situation
taken into consideration when choices were made about where to
act on the stream. In the executive summary of the proposal, the
site choice is justified by its "high public profile, its mix of public
and private land ownership" (citation, year, p. 3). The riffle was
part of a plan to generate a network of support for the stream en-
hancement project.

Community Park is very popular park. There are a number of
playing fields with stands, a lawn bowling club and a treed picnic
area included within its bounds. The trails through the wooded
section of the park are well traveled—they are a common destina-
tion for dog walkers, and it is not unusual to see a horse galloping
up the footpath in the ravine. Putting a riffle in Community Park
would ensure that a large section of the community would notice
it. And the park was well known and easily accessible, so people,
learning about the riffle, could come down for a look. With the introduction of interpretative signs, extensive publicity for the project could be generated.

Five sites in Community Park were identified as appropriate
for placing riffles (Proposal, p. 12). The site closest to the foot-
bridge crossing the creek was the one that was chosen. It was closest to the optimal habitat downstream but was also the most *visible* to a passerby. Putting the riffle at this site within Community

Park brought challenges to the project. There is no single owner of
the creek or adjacent lands. The municipality owns one side of the
ravine, a series of landowners own the other bank, and the prov-
ince controls the waterway. By choosing another location with single land ownership, the Project could have reduced the negotiation
load.

The riffle is designed to create a deep pool behind it. This pool
would have a mixed sand and gravel bottom and would extend
the trout's spawning habitat. The location of the planned pool
was also the point at which horse owners train their horses to ford
creeks—the bank's slope is shallow, the water is present and
moving, but not enough to pose a hazard to the animal. So the riffle and the pool come into direct conflict with the interests of another community of practice.

These two factors were not treated as obstacles to a goal but
rather as opportunities to engage as many "influential" members
of the community as possible. By gaining their support, the project
increased its presence in the community, through increasing the
size of their network and the strength of its connections. By having
the municipality on their side, an actor representative of thousands of residents, they significantly increased the scope of their
network and recruited a powerful ally. This was made clear when,
as we finished work on the riffle, municipal workers brought and
installed two interpretive signs near the entrance to the footpath
next to the riffle—and the activists had not even requested them!

The homeowners living along the creek represent another kind
of ally. Because they are considered "just plain folks" in the community, some of them descendants of pioneer families, their support has a strong effect on those who are not sympathetic to the
municipal government. Also, by engaging landowners, the project
again increases the number of actors in its network, stabilizing itself in the process. Since its long-term goal is to be a funding-free
organization, this stabilization is crucial.

Meagan, the project leader, sought to engage and persuade the
horse community. A "horse person" herself, she would be considered an insider within the community and could draw on her own
knowledge and familiarity of the community's culture to be an
able negotiator. The issue of the ford crossing was not a pressing

one for the riding community—it was perceived as an inconvenience rather than a big problem. By engaging the horse community rather than attempting to exclude them, the activists, with a minimal amount of conflict, enrolled another supporting actor and the people that made up its network.

Building the Riffle

Riffle construction is diagrammed by the Project using a cutaway view to highlight its different aspects—"crest," "gravel pad," "bedrock" etc. The riffle is diagrammed using triangles as if in a blueprint. The different slopes of the two sides could easily be measured and used to describe the structure—which is a triangle with its longest side on the bottom of the creek, its short, steep side upstream, and a longer side of more gradual slope on the downstream side. This description of the riffle is grounded in science and engineering practices. The experience of building the riffle was a different matter.

> The light is gray. Breath swirls out from our mouths in clouds. The damp, if you're not moving, chills you underneath all your layers of clothing. It is still, the clouds hang limp and sodden, a low gray ceiling. Cedar and fir tower above us a hundred feet or more, leaning over the creek, forming a dark evergreen canopy above us. Green and gray, these are the colors of the Pacific Northwest. We toil and try to keep warm. (Fieldnotes, 10/98)
>
> We built the riffle in layers of smooth cobble (no sharp edges to cut the fish), no less than 8 cm in diameter (palm width, no one used a ruler). After we placed a layer of rocks down, sand was used to fill in the spaces between them. Wobbling atop the slippery loose cobble, we wash the newly poured sand into the cracks with buckets half-full of stream water—careful not to get our gloves wet. It's too cold to get wet today. So washing the sand into the cracks takes a few pails filled with water. With every wash, silt erupts from the riffle in great tan plumes. It's too tricky to walk along the cobbles and get fresh water, so we draw water from the area we are washing. The water we wash with is dirty. I bend and twist as much as I can without losing my footing to reach cleaner water. (Fieldnotes, 10/98)

Here we glimpse some of the factors that inform our engagement with the riffle. Rather than angles and slopes, we have concerns of balance, keeping warm and dry, approximate "good enough" craftsmanship whose adequacy is not judged by a protractor or surveyor's tool, but by a "feel." "Looks good!" is the criterion. To judge the fitness of our efforts, we would step back and have a look—did the surface look smooth? When we stepped on them, were the cobbles firmly in place? Did our new layer of cobble present an even surface across the face of the riffle? After every few rocks we would step back and ask ourselves those questions. We answered them by feeling, using our intuitive riffle sense, and stepping literally on and pushing the riffle cobbles around in their riffle bed.

> Sand is delivered in ice cream pails, which are handy but not ideally suited to the task at hand—they are liable to break if filled too full. The cold temperature outside affects the application of the sand—we sacrifice accuracy of pouring to keep our gloves dry and our hands warm. (Fieldnotes, 10/98)

By participating in this act of construction we move away from the authoritative certainty of mean-bankfall width and slope and learn about a practical sense of placing cobbles and filling the interstices with sand and keeping balance.

> We used the shovels to empty the pickup truck of its load of sand, creating piles on the ground, and then used the shovels to fill the ice cream buckets we used to carry the sand to the riffle. We also used first brooms, then our booted feet to clean the sand out from between the ridges on the floor of the box (back of the pickup truck). Meagan had in the box an 8-inch galvanized nail, which she would run along the crack between the box and the tailgate to clear out the sand that had been packed in there. The tailgate wouldn't close. Meagan told me to just slam it, but it still wouldn't latch on to the truck. I discovered that there were pebbles in the two indentations where the clamps, which clamped onto the body of the truck lay. Looking down on the ground, I found some small sticks and used them to extract pebbles from the recess that held the clamps. (Fieldnotes, 10/98)

In this case, multiple tools and strategies are necessary to get a relatively simple task done. Boots, shovels, nails and twigs were recruited to rid the truck of sand. Their use was highly context dependent—dependent on the progression of the job, where the sand was going and what space was being cleared.

> Later in the day, I hung Fran's keys on her new broom. She had tossed
> her keys amongst the roots of a cedar tree, and with all the activity
> swirling around, I was worried that they might get covered up or that she
> would forget where she'd tossed them. I was also concerned that she
> might forget to take her new broom home, as it was just tossed off to the
> side, and she hadn't brought it (Meagan had just purchased it to replace
> one crushed by boulders). By hanging the keys on the broom, the keys
> were off the ground and out of harm's way, and she wouldn't forget the
> broom. (Fieldnotes, 10/98)

In this vignette, the boundary between the mental and the
physical worlds become fuzzy. The keys hanging on the broom
could be thought to be doing mental work, that is, they are reminding Fran to not forget the broom and protecting the keys from
getting accidentally covered by leaves or otherwise lost. These are
skills we all use in practice, whether it be running an electrophore-
sis gel or arranging the screws and nuts we remove from our bicy-
cle when we fix it.

In much of these activities, we observe an enactment of *sens
pratique* (practical sense, translated as "logic of practice") which
is characterized by a practical coherence that differs from the coherence of detemporalized, objectifying, scientific coherence:

> Their unity and their regularities, and on the other, their fuzziness and
> their irregularities and even incoherences, which are both equally neces-
> sary, being inscribed in the logic of their genesis and functioning—to the
> fact that they are the product of practices that can fulfil their practical
> functions only in so far as they implement, in the practical state, princi-
> ples that are not only coherent—that is, capable of generating practices
> that are both intrinsically coherent and compatible with the objective
> conditions—but also practical, in the sense of convenient, that is, easy to
> master and use, because they obey a 'poor' and economical logic,
> whereby no more logic is mobilized than is required by the needs of prac-
> tice. (Bourdieu, 1990, pp. 86–87)

Bourdieu criticizes "objective" theoretical constructs with their
"forced synchronization of the successive, fictitious totalization,
neutralization of functions, substitution of the system of products
for the system of principles of production, etc." (p. 86). Because
of their conventions and structure (some of which are mentioned
above), they lose their ability to describe how things are done.

This problem is relevant to science education because above all, science is an activity, and it is often presented as a collection of objective facts.

The experience of riffle building is shot through with this logic, whether it be placing cobbles where the riffle demands it, or the urgency to complete felt by the riffle builders as the day progressed, the light faded, and the participants got more and more chilled. This is important because it is the logic of not only life in the lab, but also of how things are done in daily practice. Washing dishes—rinse or not, fill the sink, dry with towel or rack—or cooking a meal in the middle of a busy day—what is wilted, what is fresh, how much time is there left, what people have a hankering for. During the day's work on the riffle, Stuart was continually challenged with situations which demanded his technical improvisation or judgement, and the feedback about the effectiveness of his choices was rapid and concrete (for example, did the pickup tailgate close?). Through engagement fraught with these challenges, students will learn about the art of doing, which is at the heart of science.

Through this intimate act of building and knowing, the participants learn about stream ecology. They see and hear the changes that the riffle makes in the stream. "Spawning habitat" also becomes a clear reflecting pool, "increased oxygenation and reduced stream energy" also becomes a delightful burbling brook, where once a silty trickle slouched towards the sea. Oxygenation becomes an aesthetic value, capable of binding people in a group.

> The riffle had been working its magic, and overnight a pool had built up behind it; the water was probably about three feet in the deepest area. Certainly well above our gum boots. A creek I remember as a trickle[5] (when I helped in the surveying) was now a lovely pond, cascading with a delightful gurgle over a bed of rocks. The creek seemed more alive than it had been. The sound of the water through the riffle was delightful, and

---

5. This section of the stream previous to the building of the riffle was a tiny trickle of a stream limping its way through barren black mudflats, a collage of rotted leaves and discarded twigs. Fine black silt from the winter floods had covered everything. Nothing grew. The black earthscape was punctuated only by the brilliance of the occasional beer can or potato chip bag.

we all reflected on how the riffle seemed to make the murky water clean
and clear. The sound itself seemed to increase the stream's vitality. The
deep clear pool was serene and somehow seemed fertile, now a rich spot
for life (many lives) to spawn and thrive. [Fieldnotes, 10/98]

Through our work, the members of the group learned to trust
each other, and through our shared aesthetic satisfaction, rein-
forced by many visitors' comments, they came to regard each
other in sort of a glowing light. Stuart learned a little about Karen,
a water technician, and her down-to-earth, "hands-on-working"
values. He had fun playing within his happy relationship with
Meagan. He met and chatted with Fran, a woman whose sincerity,
love of the land and personal industry he came to admire.

During the course of the day, many people came by to visit the
riffle site and therefore became part of the network that stabilized
the riffle within the community.

(1) A photographer from the *Times Columnist* (local newspaper) visited.
She spent about ten minutes of our time taking pictures of Karen (water
technician) and myself picking up the rock, putting down the rock, talking
to each other, picking up the rock, putting down the rock. We filled her in
a little about what it was we were doing (necessary for the caption).

(2) Tom from DFO showed up. He had done a lot of stream restoration work, up on Rocky Creek in nearby Northtown. They had been
working that creek for about 15 years, bringing back both salmon and
trout. Tom had a look at the riffle and the riffle pool, and gave us some
suggestions about improvement.

(3) Spring came by. He is a sole proprietor of a GIS/GPS company.
Things are going a little slow in his business. He has volunteered to do
the GIS mapping of the area.

(4) Marie dropped by for a couple hours in the morning. She is the
wife of a heavy equipment dealer/repairman. She lives nearby. She
helped us out with the rock lifting. Her daughter Yvonne was also there,
and stayed with us after her mom went home.

(5) Gordon came by. He is an old-time resident of the peninsula, is in
his seventies and though he walks with a cane, is alive and open and
sprightly looking. He was telling all of us stories about the creek and the
surrounding land.

(6) Bonnie dropped by for a few minutes, and spent about twenty
minutes talking with Fran about the project. They were talking about
councilors and civil bureaucrats and what was going on in local politics.

(7) Two young girls on horses trotted by as I hauled rocks into the wheelbarrow on the road. Signage had been placed on the road, indicating the creek improvement project—upon reading the sign, one of them said to me, "Does this mean we can go fishing in the creek?" "Maybe," I enigmatically replied, and they galloped up the hill.

(8) Dogs loved the riffle pond. Every retriever who went for a walk that day jumped in. We could only hope we had made some good trout habitat, but we were sure we had created good dog habitat.

(9) Lee, from the local newspaper dropped by. Meagan and I posed for pictures, squatting in the pond.

(10) Allen also dropped by. He and I and Loreen surveyed the creek bottom a few weeks ago. He and Meagan argued about some points regarding the riffle, some technicalities.

(11) Martha, 50-ish, came by and lifted rocks for an hour or so just before lunch. I think she just came by to help. Seems she lived nearby.

(12) In the afternoon, two steering committee members dropped by. Meagan had neglected to inform them the regular meeting was cancelled. So they decided to have a look at the riffle which they had spent so many hours in planning.

(13) Two other guys (techy types) dropped by to check on the progress. Meagan spent 10–15 minutes discussing the riffle, pool, and plans with them.

(14) Another woman, whose name I forget, but who works at the nearby Federal Marine Research Institute in toxicology dropped by to see how things were going. We chatted about the lovely nature of lab jobs.

(15) Karen is a water technician hired by Oceanside farm. We shoveled sand/gravel out of the back of Meagan's pickup truck all morning. She left after lunch, around two o'clock. (Fieldnotes, 10/98)

In this passage we see the diversity of interests and connections to the stream and the riffle that converge on our construction project. The newspaper photographer is an actor makes hundreds of thousands of readers witness events in a virtual way. Three months later I am still reminded by acquaintances of my photo appearing in the daily paper.

Tom and the two unknown techies come by and spend close to half an hour discussing riffle particulars and probable consequences of certain actions: whether or not to in-fill under cut root caps. Are they likely to provide useful habitat, or will they likely be further eroded, and result in the loss of the tree? What would happen if we put a large log in the pool to reduce the force of the current as it approaches the riffle? Where would the water be directed, what impact would it have on the surrounding banks? Through these informal exchanges we learn about the active riffle in the actual stream. They reassure us that the riffle is looking good and is doing what it should be doing.

We meet others who have been working on the Henderson Creek project and come to know them in a context where it is clear that I support the project, and this gives me common ground with them. This is important for me as a newcomer and helps my being accepted into the group. It is also fun. Because we are building something tangible, enacting our plans, finally doing something towards our long-term goal of restoring the stream to "health," the spirits are high, and there is lots of opportunity for storytelling and humor which bring us closer together. We met many members of the community, and their encouragement and admiration and questions and interest helped me learn about the social environment within which Stuart was participating. The stories told by the people who dropped by the riffle while going for a walk were an education in the community's concerns and history.

The participant in such a situation learns about the community in which he or she is embedded, the history, priorities, mistakes, and conflicts that have happened and continue to happen over time. We get a sense of participation within a greater whole, a sense of knowing that cannot be separated from the accents, inflection and pacing of the conversations we hear and later carry on with others in the same locale. Rather than being some abstract concept, it is a process of absorbing the practices and concerns of a community at large, while having a definite place within the community (the creek restorationists) that helps to position us with respect to our conversations with others and lends us an identity. This is especially important for children, who are often seen as irrelevant in terms of contribution to the community, and may give them identity as active agents.

> Meagan told the story of how this riffle was achieved politically. First, she solicited support from DFO (federal Department of Fisheries and Oceans), then MELP (provincial Ministry of Environment, Lands and Parks) through her contacts. She worked a lot with the people who owned the banks on one side of the creek and persuaded them to support the riffle. Then she presented council with the project and all the commu-

nity support. It was instantly approved on Monday, and here we are, finishing it on Friday.

But later, Meagan told us the second half of the story. And that is that she did not have explicit approval from MELP. She was trying to float the application through on a rapid approval technicality "section 9" which is normally reserved for government projects. Although she was told her project would have no problem getting quick approval, the Ministry wouldn't grant it, even by Tuesday evening. Meagan wanted the riffle in very soon, (it was mid-October) before the winter rains came and made the work impossible. So with all the volunteers ready to go on Wednesday, but without MELP approval, Meagan decided to go for it anyway. She gambled (I am guessing it is bad to break provincial watercourse law), hoping her personal connection in MELP and technical connection to DFO would combine to allow her to receive fast-track approval. Disregarding bureaucratic imperative, she forged ahead and began the work, only to find out a bureaucrat had misinterpreted her application—the bureaucrat thinking that the federal research institute was a private foundation instead of a federal research institute. (Fieldnotes, 10/98)

Once again, we learn the crucial role that networks play in enacting science—both the informal network of residents Meagan recruited and the structured network of provincial civil servants charged with regulating watercourses. Without Meagan's efforts at bringing residents "on board," the riffle project would never have left the report and entered the Creek. Through animal health, bio safety, hazardous waste and other committees, limits are put on what scientists can study—their world is constrained by the demands of the bureaucracy. In this situation, the bureaucracy misfired, withholding approval and putting the project at risk. Meagan's sense for the discourse of the regulations and regulators supported her in her decision to ignore the blockade set up.

Gordon, an old-time resident of the peninsula and the owner of a farm downstream from the park, was telling us stories. How people used to get up early before work, catch their limit of salmon by 8:01, and then head off to work. Now a salmon is rare sight in the inlet. He told us how there used to be schools and schools of herring churning the waters, but not anymore. We learned how there was a deep pool at this site, one where the children used to swim. The municipality had it filled in for safety reasons during the fifties. (Fieldnotes, 10/98)

Through storytelling, Gordon informed those present of the natural history of the region. These are the kinds of nature lessons that no textbook or video can give. We learn from someone while we are present in a space, about that space. We can ask questions about what we are interested in, or for more stories. We learn about the past, and this gives us ties, through our vision and ability, to the future. It engages us in the flow of time.

Gordon's stories instilled a deep sense of sadness and anger within Stuart at how arrogantly, ignorantly and violently we had treated our natural surroundings. His stories strengthened Stuart's resolve to keep this restoration work going and to educate myself, to participate more fully, and thereby be able to educate others. Maybe one day, Stuart's work would contribute to a return of some herring to the inlet, or maybe salmon.

In this chapter, the story of the riffle is bounded by the constraints of written publication. In the life of the community, there was no clear beginning to the riffle story (though we can always construct some criteria according to which we can define a beginning). The story really continues because the riffle has become part of the life of the community (including the grade-7 students whom we present in Roth and Lee, this book).

Is This Literacy?

Upon reading an early version of this paper, a colleague remarked that what we were describing was interesting, but why call it science education? This is an important question to answer considering that that this experience looks very different from the one we are used to: students poring over textbooks, taking notes, doing simple experiments designed to demonstrate a concept like "pH." We scaffold our analysis on Fourez's (1997) recent work on scientific/technical literacy. In it, he suggests that scientific literacy consists of the right use of a variety of scientific resources: specialists, black boxes, simple models, interdisciplinary models, metaphors, standardized scientific knowledge, translations, and knowledge and decisions. How many of his criteria did we fulfill? Could this activity be justifiably claimed as strong scientific literacy? What would legitimate peripheral participants learn on a day at the riffle?

Right use of specialists: From its inception, specialists' advice guided the project. Stream biologists advised and enacted the initial surveys; as mentioned in the case study, Tom gave important advice about what to do and referred Meagan to another specialist, Chris, her co-worker. The project has had extensive support from its technical advising committee.

Right use of black boxes: This is the ability to judge when not to open a phenomenon up to analysis, but rather to just let it do its thing. That is, we do not need to know how a computer keyboard informs the CPU of the letters we are pushing, we are happy to use it as a black box. There were not many technological black boxes used in the riffle construction, as it did not rely on the use of high technology. But the empirical evidence, which resulted in the decision to build the riffle, could be considered a black box; as we did not question the need for a riffle nor the theory that suggested what it would do for the habitat.

Right use of simple models: This is knowing when a situation needs to be explained theoretically, for example, "what model would be appropriate to work out when it is convenient to pull and when to push a wheelbarrow?" (Fourez, 1997) In this case, simple models about water flow and stream behavior were continuously modified as we discussed rock size to place in the riffle, effects of sand fill, where silt would accumulate, etc.

Right use of interdisciplinary models: This notion refers to the invention within the context of a specific project, of an adequate model—fairly simple but using knowledge stemming from various disciplines as well as from the know-how of everyday life (Fourez, 1997). There was not much call for these types of models in the project, as the construction of the riffle did not require the expertise of specialists from different disciplines.

Right use of metaphors: Metaphors of relationship to land, of the nature of the stream and so on permeated the discussions all day long.

Right use of standardized knowledge (scientific disciplines): This means that students have to be inducted into established views and methods, that is, those that have been successful and without which it would be practically impossible to communicate within a scientific and technical society (Fourez, 1997). The riffle itself and the justifications for building it were embedded within the standardized knowledge of restoration ecology. In this case the standardized discourse was used "economically" (Bourdieu, 1990); we weren't trying to "learn ecology," we were learning enough to get the job done. The practice of ecological restoration is grounded in the Western scientific notions of prediction, control and experimentation. By participating with the group, enacting change and believing that our informed actions are doing good, we live out a master narrative of Western science. (On the contrast between Western science and local/Aboriginal science, see Aiken-head, this volume.)

The right use of translations: This is the skill of translating standardized knowledge into representation of everyday life—and vice versa, analyzing our everyday life situation in terms of standardized knowledge. Trout were the theoretical constructs of a successful riffle and also physical beings we sought to discover resting in the pool. Logs and bankside trees became "large woody debris" and undercut bank became "habitat." The talk around the riffle building was rich with these translations.

The ability to contrast the understanding of a technology with the understanding of its scientific principles: This refers to the difference between understanding how to use a fax machine, what it is useful for as opposed to e-mail or telephoning (technological understanding) and the scientific principles behind its operation. There was not much of this discourse at the riffle site, as the technology involved was simple and non-problematic.

Right use of knowledge and decisions pertains to how we teach young people to relate scientific and technological knowledge to ethical and political decisions (Fourez, 1997). At the riffle site our discussions continuously turned around this topic: what the creek's problems were, who was responsible for them, what was it about the political economic climate that encouraged the problems. After a day at the riffle, students would be well versed in many versions of how science and decisions have been related over the years.

The experience of building the riffle was strong in terms of the following categories: specialists, simple models, metaphors, translations, knowledge and decisions. It was moderate in terms of black boxes, standardized knowledge and not a great learning area for interdisciplinary models or the distinction between science and technology. This reflects the relatively simple technological nature of the task, which did not require much use of instrumentation or drawing specialists from a wide variety of disciplines.

Overall then, with 6 of Fourez's 10 categories strongly represented in the riffle-building experience, we are justified in our claim that this is a good way to teach students practices that would result in scientific literacy.

Practice

Throughout the history of the riffle construction, the practice of this science-based activity is significantly different from that performed in classrooms. The science was purposeful; it had a strong goal that it supported which had nothing to do with the advancement of science per se or "getting it right" to please an authority. The goal of enhanced watershed and stream health determined the type of the science carried out in the project. And in return the results of the science shaped the actions and discourse of the activists.

The science enacted in this case study is aptly described by Bourdieu's "economy of logic" (Bourdieu, 1997)—no grand narratives were constructed and painstakingly checked and double-checked. Science was used as a tool to help people determine what was the appropriate action to take. People's activity while building the riffle was directed toward working in community to build a structure that functions properly. There were no authorities withholding marks for not doing it fast enough or in exactly the right fashion or not understanding the underlying concepts—as during the Initiation, Response, Evaluation sequences of teacher-student interaction (Poole, 1994) that reify notions of science as a body of fact. Peoples' activities were guided by what needed to be done next in the construction sequence. The fading of the daylight and threat of oncoming winter rains gave the project its urgency instead of an impending bell signifying an authority's decision about what to do next. Thus, the experience of time while working on the riffle was closer to that of the practice of both professional and everyday science-based activities. In these activities, urgency is embedded in daily rhythms (meals, bed time) and activities determined by the next step necessary for the successful completion of an experiment, design project, or course of treatment. The activities embodied the fuzzy and economical logic described by Bourdieu (1990), and they were rich in improvisation, adaptation and embodied knowledge. Based on this view of the nature of practice, participants gained a legitimate experience of scientific practice and of practice in general.

Connections

One of the most outstanding aspects of the science enacted by the activists was its connectedness to its community. Geographically, technically, and discursively the science was profoundly influenced by its relationships. This science was not confined within four walls. It traveled to multiple locations: municipal council, landowners' living rooms, Henderson Creek, committee meetings, and so forth. Through its travels it changed its meaning and its form, from an inconvenience discussed between individuals to the horse community, to the formal presentation of data in a technical report to the municipality to a group of people building a riffle. Each site and corresponding community contributed to and shaped the project.

The science as enacted was determined entirely by the community. Its land-use practices created the "problems" to be solved; the funding available determined the extent of work that could be done, and the people who participated in and supported the project allowed it to happen. However, science provided a legitimate description of the problems, and justified the group's actions. Inasmuch as scientific description is likely to convince people that action is needed, science defines the problems and guides the community's action.

As we exhaustively demonstrated earlier, the riffle construction was tightly woven into the community. Many people came by to help, visit or advise. Through these stories, we become more deeply rooted in our community, absorbing its language and rhythm, priorities, and history. As the participants come to know a place—its people, its history and politics—a sense of belonging develops which is a crucial ingredient for being able to inhabit a place as a dwelling (Heidegger, 1977). Sensitivity to biogeographical surroundings develops a feel for shared cultural history and the forces that shape our habits and assumptions. We learn about that part of us which is the land—its needs for proper treatment, its native tendencies and become intimate with it by sticking our hands in it and shaping it. Therefore, the ultimate learning that goes with this type of activity is one of a relation of Self to Other, a becoming in the world.

By participating in the project, even for a day while building a riffle, one becomes an increasingly empowered social actor. Much of the talk during the day at the riffle deconstructed taken-for-granted practices (ditches, storm sewers, etc.), and opened them to critique. Through participating in this discourse, we begin to analyze and critically think about our place rather than passively accept whatever some authority decrees.

In Lieu of a Conclusion

The kind of science we have described in this chapter is quite different from that being currently taught in schools today. We do not mean to say "this is the only way," but rather to provide a different framework with which to understand scientific literacy and show some opportunities available to them to help their students develop meaningful relationships with scientific practice. We flesh out beginnings of a different science education in another chapter (Roth & Lee, this volume). Although we are still far from where we want to go with students, we are hopeful that such beginnings can change the tide. We want to afford students the opportunity to contribute knowledge to the community at large through their engagement. And we want to get more students to participate not in the science of scientists, but in science as it permeates everyday activities in our communities. We envision a school science that includes purposeful projects, multiple sites, a wide range of literacy skills, and a focus on practice. We see beyond the "cold" science of facts stripped of politics and passion to a science densely woven into the interests that shape it and in return are shaped by it. We see science education not as a preparation for a future life, but as an active legitimate participation in a community.

Acknowledgments

This research was supported by grants 410-96-0681 and 410-99-0021 from the Social Sciences and Humanities Research Council of Canada. The views represented in the chapter are our own.

References

Bourdieu, P. (1990). *The logic of practice*. Cambridge, England: Polity Press.

Brookhart Costa, V. (1993). School science as a rite of passage: A new frame for familiar problems. *Journal of Research in Science Teaching, 30*, 649–668.

Eisenhart, M. A., & Finkel, E. (1998). *Women's science: Learning and succeeding from the margins*. Chicago: University of Chicago Press.

Epstein, S. (1995). The construction of lay expertise: AIDS activism and the forging of credibility in the reform of clinical trials. *Science, Technology, & Human Values, 20*, 408–437.

Epstein, S. (1997). Activism, drug regulation, and the politics of therapeutic evaluation in the AIDS era: A case study of ddC and the 'Surrogate Markers' debate. *Social Studies of Science, 27*, 691–726.

Fourez, G. (1997). Scientific and technological literacy as a social practice. *Social Studies of Science, 27,* 903–936.

Fuller, S. (1997). *Science.* Buckingham, England: Open University Press.

Gieryn, T. (1996). Policing STS: A boundary-work souvenir from the Smithsonian exhibition on "Science in American Life." *Science, Technology, & Human Values, 21,* 100–115.

Haraway, D. (1995). Situated knowledges: The science question in feminism and the privilege of partial perspective. In A. Feenberg & A. Hannay (Eds.), *Technology and the politics of knowledge* (pp. 175–194). Bloomington: Indiana University Press.

Hazen, R. M., & Trefil, J. (1991). *Science matters: Achieving scientific literacy.* New York: Doubleday.

Heidegger, M. (1977). *Sein und zeit* [Being and time]. Tübingen, Germany: Max Niemeyer. (English translation consulted by J. Stambaugh, State University of New York Press, 1996.)

Knorr-Cetina, K. D. (1981). *The manufacture of knowledge: An essay on the constructivist and contextual nature of science.* Oxford: Pergamon Press.

Knorr-Cetina, K. D. (1992). The couch, the cathedral, and the laboratory: On the relationship between experiment and laboratory in science. In A. Pickering (Ed.), *Science as practice and culture* (pp. 113–138). Chicago: University of Chicago Press.

Latour, B., & Woolgar, S. (1986). *Laboratory life: The social construction of scientific facts.* Princeton, NJ: Princeton University Press.

Latour, B. (1993a). *La clef de Berlin et autres leçons d'un amateur de sciences.* Paris: Éditions la Découverte.

Latour, B. (1993b). *We have never been modern.* Cambridge, MA: Harvard University Press.

Law, J. (1994). *Organizing modernity.* Oxford, UK: Blackwell.

Lemke, J. L. (1990). *Talking science: Language, learning and values.* Norwood, NJ: Ablex Publishing.

Longino, H. E. (1995). Knowledge, bodies, and values: Reproductive technologies and their scientific context. In A. Feenberg & A. Hannay (Eds.), *Technology and the polititics of knowledge* (pp. 195–210). Bloomington: Indiana University Press.

Pickering, A. (1995). *The mangle of practice: Time, agency, & science.* Chicago: University of Chicago.

Poole, D. (1994). Routine testing practices and the linguistic construction of knowledge. *Cognition and Instruction, 12*, 125–150.

Roth, W.-M. (1998). *Designing communities.* Dordrecht, Netherlands: Kluwer Academic Publishing.

Roth, W.-M., Boutonné, S., McRobbie, C., & Lucas, K. B. (1999). One class, many worlds. *International Journal of Science Education, 21*, 59–75.

Roth, W.-M., & Bowen, G. M. (1995). Knowing and interacting: A study of culture, practices, and resources in a Grade 8 open-inquiry science classroom guided by a cognitive apprenticeship metaphor. *Cognition and Instruction, 13*, 73–128.

Roth, W.-M., & McGinn, M. K. (1997). Deinstitutionalizing school science: Implications of a strong view of situated cognition. *Research in Science Education, 27*, 497–513.

Roth, W.-M., & McGinn, M. K. (1998). >unDELETE science education: /lives/work/voices. *Journal of Research in Science Teaching, 35*, 399–421.

Roth, W.-M., McGinn, M. K., Woszczyna, C., & Boutonné, S. (1999). Differential participation during science conversations: The interaction of focal artifacts, social configuration, and physical arrangements. *The Journal of the Learning Sciences, 8*, 293–347.

Roth, W.-M., McRobbie, C., Lucas, K. B., & Boutonné, S. (1997). The local production of order in traditional science laboratories: A phenomenological analysis. *Learning and Instruction, 7*, 107–136.

Tobin, K. (1990). Research on science laboratory activities: In pursuit of better questions and answers to improve learning. *School Science and Mathematics, 90*, 403–418.

Tobin, K., & Gallagher, J. J. (1987). What happens in high school science classrooms? *Journal of Curriculum Studies, 19*, 549–560.

Winograd, T. (Ed.). (1996). *Bringing design to software.* New York: ACM Press.

4

Breaking the Spell:
Science Education for a Free Society

Wolff-Michael Roth & Stuart Lee

In the past, science education has been under the spell of scientists. Science educators and science teachers have subserviently done the job that scientists wanted them to: reproducing a sorting system and existing inequalities that shake many students out of science and technology related career paths (Roth & McGinn, 1998). This spell is especially evident when we look at science from another domain interested in how scientists work and how they construct new knowledge: science studies. Sociologists, anthropologists, feminists, ethnomethodologists and others interested in science have shown both the ordinariness of scientific practices and, at the same time, the rhetoric and politics of scientific communities that attempt to reserve themselves special places in society and cull a maximum of financial resources from society (Fuller, 1997). That is, not only does science education as it is practiced shake people out of the system but those who succeed in the system become a class of high priests with enormous influence on the way resources are deployed in our society. (Of course, scientists themselves are not just powerful agents, but are themselves bound up in complex networks of other actors and forces that constrain what any single or group of scientists can achieve.) We therefore raise some serious questions such as, "Is this the kind of science we want to teach?," "Do we want to continue to use science education as a career selection mechanism or do we want science for all?," and "What would be an appropriate science that takes seriously the word 'for *all*.'"

Our agenda, "Science education as/for sociopolitical action" constitutes a further development of arguments made earlier (e.g., McGinn & Roth, 1999; Roth, 1998b; Roth & McGinn, 1997a). Rather than thinking about science education in terms of scientific knowledge worthwhile on its own, we propose to think in terms of worthwhile purposes which we, as science educators and science teachers, pursue with the beings (children, students) in our care. These purposes cannot be argued from within science but are related to the values we hold. But values are matters of choice rather than logic, and there is nothing preventing us, as a society, to value a philosophy of wisdom in which the discourses of music, literature, drama, politics, science, religion, and philosophy are treated at the same level (e.g., Maxwell, 1992). Science then becomes an exemplar of solidarity rather than one of rationality and objectivity. Such a new and differently conceived science will have implications for the ways we teach the subject; for example, we might teach an integrated subject where science is but one of the many threads of the weft that cross the warp of topics and activities (e.g., Lee & Roth, this book). A reconceived science education may then provide a context for educating future generations in a society free from the spell of science but which nevertheless draws on science to deal with the ecological and technological challenges it will face. In this reconceived science education, literacy is no longer defined in terms of scientists' science (e.g., Hazen & Trefil, 1991). Scientific literacy might then be defined in terms of knowing about the right use of specialists, scientific black boxes, simple models, interdisciplinary models, metaphors, standardized knowledge, and translations, negotiations, and knowledge transfer (Fourez, 1997).

To be pedagogically and societally viable, science education has to change its face. Currently, science education has both sorting and propaedeutic functions (e.g., Brookhart Costa, 1993; Roth & McGinn, 1998).[1] The standard reference for what is to be taught

---

1. Propaedeutic is a term much more frequently used in European (French, German) education and literally means 'teaching before hand,' that is, teaching as an intro-

is always that of a modernist science as a body of facts, laws, and theories as presented in science textbooks or in curriculum documents. This focus on "real" science (as defined by hardcore scientists) has not changed with recent shifts in the epistemology underlying curriculum reform. Thus, with the emergence of constructivist epistemology, we have seen a change in the way educators (and some teachers) conceive of learning, but what we ask students to learn, and which is the referent during testing, is still the same "cold" science. Even our own efforts in curriculum reform did not escape from the spell of science (e.g., Roth & McGinn, 1997b; Roth, McGinn, & Bowen, 1996). That is, we had advocated science education to be modeled on a view of science as it emerged from the science studies literature. We framed science in terms of a social constructivist metaphor of the emergence of individual and collective knowing and learning, and the relationship between individual and communal knowing.

In this chapter, we break with this tradition and thereby deconstruct[2] the spell of scientism over science education. We presuppose that an adaptive, postmodern society needs a different kind of science education, a *science education as/for sociopolitical action* that has as central goal the betterment of the world.

Epistemology and Science: Indoors and Outdoors

In this section, we review two major bodies of research that both question modernity's concepts of what school science might be. In

---

duction for future things come. School science education, if viewed as a propaedeutic effort, is therefore concerned with training (filtering) for the next higher levels and would therefore essentially a job selection and job preparation mechanism. In schools, this function is evident in teachers' continuous concern for covering some curriculum to prepare students for the next grade or for university entry. Grades and the associated actor network which trades in grades and associated currencies (e.g., sports, extra-curricular activities) are used to make such selections, often along lines of gender, race, or socioeconomic status (Roth & McGinn, 1998).

2. As Derrida (e.g., 1988), we use "deconstruction" not as (senseless) destruction, but in the Heideggerian senses of Destruktion (dismantling) and Abbau (to take apart an edifice).

the sections on situated cognition, we suggest that there is a fundamental discontinuity between what and how people know mathematics and science in and outside schools. In the second section, we review the literature on activism that shows that groups with no scientific training in schools were able to define, in the political and legal arenas, what science is and how it should be conducted.

Research on everyday cognition shows that there is more to knowing than facts and skills (e.g., Lave, 1988). A much more appropriate ontology of everyday cognition includes standard practices, material resources, linguistic resources, breakdowns, and ongoing concerns. A comparison of examples from each of these categories with traditional science education shows that it is concerned only with some, linguistic (but decontextualized) resources and perhaps some standard (but so highly routinized as to be laughable) practices. Students never get to know about the ongoing concerns and never engage linguistic resources in the service of some authentic goal to be achieved, and never get to engage in the practices of scientists or any other science-related field.

Transparency of Knowledge and World

Over the past decade, it has become increasingly clear that theories which conceive of knowledge as procedural and factual propositions residing somewhere in the brain, physically instantiated in symbolic form, are clearly inappropriate (e.g., Hutchins, 1995; Lave, 1988). Recent research increasingly shows that it is more appropriate to think of knowledge not as something static; rather, we should be thinking of knowledge as something enacted, that is, knowing is always revealed in action (Bourdieu, 1997; Orr, 1998). But because actions occurs in some physical and social context, all knowing in action is irremediably situated. Knowing is what we instantiate in our interactions with our world; knowing does not consist in memorized factual statements which are disconnected from everything else we do.

Much of what we know is not salient in our daily actions. Thus, when we use familiar tools, these become transparent to our activities:

> Michael: When I type my parts of this chapter, I do not have to think about the keyboard and its individual keys, but I think about ideas, texts, and arguments. In fact, most of what I know about computers—though colleagues and students call me a computer freak—I cannot tell in words but have to enact by doing it with the computer. I do not know by heart which window to pull to get my word processor to change some font. But when I use the word processor, I pull the appropriate window or, by means of a scan with the mouse, get to the appropriate window to be pulled. Thus, much of what I know resides in the background and is presupposed in the activity itself. It is only when things do not go the way I expect them to go that I begin to inquire and make aspects of my world salient to reflect on them. This reflection needs representation, and it is at best at that point that we engage in the kind of rational activities that traditional psychology is all about.

In the way we described our own activities of using a computer, scientists do a lot of things without ever having to consciously think about them. Much of what they know is hidden in their common sense assumptions about how the world works. Much of this knowledge has been appropriated through years of working with others on similar problems, in similar places, by using the tools everyone else uses, by sharing the kind of stories everyone else shares, by trying to present the results of one's work to different kind of audiences. In a recent two-year ethnography of a scientist at work we showed that much of her science was not filled with daily excitement (Roth & Bowen, 1999, in press). Rather, it was filled with days of searching for lizards, skinks, rubber boas, and garter snakes. Mosquitoes attacked in the morning as we hiked to the slopes and in the afternoon when we returned to the car. In the scorching heat of the valley, often close to 40° Celsius, the rocks we had to turn to find the animals had heated to over 60° Celsius. Some of us brought up to 4 liters of water to replenish what we sweat out climbing the steep slopes. And in the end, after 6 hours our foursome sometimes returned with only one or two animals. And yet, our data reveal that during these hours seemingly without anything happening, our scientist developed a deep and intricate understanding which underlie the complex reports she communicated, and much of the graphical representations she used in the process (Roth, Masciotra, & Bowen, in press).

Opaqueness of School Knowledge

In schools, however, we attempt to teach as if knowledge was a commodity that can be transferred from one person to another independently of the social and material contexts of its use. Constructivist metaphors seemed to have changed little in the way we conceive of and teach school science. Subject matter is still taught independently from students' interests, the contextual concerns of the subject matter in action, and the characteristic breakdowns that people have to cope with in their everyday activities. Educators claim that learning science is good for its own sake. This claim, however, is highly suspect and entirely questionable. Take the following example from mathematics. A standard activity around grade 9 is the factoring of polynomials, that is, finding that something like $x^2 - x - 6$ can be expressed as $(x - 3) * (x + 2)$. We have yet to see a rational proof that knowing to factor a polynomial helps anyone become better at everyday mathematical activities. These may include shopping for groceries, completing an income tax form, making a profit from stock investment, keeping track of baseball statistics, or calculating recipes for a different number of people. Or these may include any other mathematical activity people engage in as part of their professional or private lives. Furthermore, in one project, we counted about 25 different ways of finding a solution to such factoring problems, and yet, depending on the current topic, there is only one legitimized answer possible at any given time.[3] In his doctoral study on mathematical reasoning in college students, Roth showed that more than 60% of college students did not use ratio and proportion for

---

3. In the same vein, some of Roth's high school physics students were penalized once they went to university because they used advanced matrix algebra to solve systems of linear equations on their handheld calculators rather than going algebraically, step by step, solving for one variable at a time.

solving problems where these might come in handy, for example, in determining the concentrations of two frozen orange juice-water mixtures (Roth & Milkent, 1991). In a similar way, students are asked to memorize—for few understand them—Newton's Laws or regurgitate facts about atomic theory without ever having any opportunity to engage in a public discourse where knowing, that is, being able to talk about, atoms has any use.

Detailed analyses of mathematical performance in out-of-school situations have shown that years of schooling are not related to everyday mathematical competence on the street (Lave, 1988; Saxe, 1991; Scribner, 1986). These analyses included situations where people made best buys in supermarkets, worked as street corner bookies, or earned a living as a child street vendor of candy. Memorizing more mathematics-related information in school and doing piles of worksheets one topic at a time does not lead to higher performance in everyday mathematical practices. That is, there exists a deep rupture and alienation between practical everyday knowledge and school knowledge.[4] Many thinkers have long argued on various grounds—including pragmatist (e.g., Dewey, 1933), phenomenological (e.g., Heidegger, 1977), and Marxist grounds (e.g., Bourdieu, 1997)—that much of what we know results from our experience of acting in the world, and in terms of the community of which we are necessarily part.

Situated Cognition and Constructivism

There are readers who may want to ask us about constructivism and our ideas about situated knowing. Here, we have to say that constructivism over-emphasizes the rational aspects of how we come to know. Constructivism easily conjures up the image of each individual building knowledge from little components much like a builder puts one stone on another or one log onto another to construct some dwelling. We find the example of learning words

---

4. We hesitate to call "knowledge" that which students memorize and supply to standard school questions as answers, but which is different from the knowing which we enact whenever it is needed.

from dictionaries interesting in many respects (Brown, Collins, & Duguid, 1989). In this example from research, the authors show how some children integrate a new word, but do so in inappropriate ways so that when they use the word it does not follow the conventions. Yet children learn words in everyday settings at an enormous rate. It is when they come to school and are to learn words out of context that they begin to fail. Thus, whatever constructing these children did, attending to the words at hand, it did not help them much in learning new words. On the other hand, much of what and how we know is actually not acquired in formal situations or by focusing our attention to it. We learn our mother tongue largely without formal instruction prior to kindergarten; we even know to distinguish structurally correct from incorrect sentences, and know how we can express something without ever knowing anything about grammar. We do so, because this is the way our social and physical world is. This world is always and already there, from the beginning, and always already shot through with meaning. We embody this knowledge even without attending to it and because, as physical and social bodies in a physical and social world, we have incorporated the structures of this world. It is for this reason that constructivism has become a referent of decreased importance in our thinking about knowing and learning.

Situated cognition approaches decenter traditional cognitive research and constructivism and focus explicitly on participation in activities as these are shaped by individuals-acting-in-settings. That is, the patterns and structures we come to experience in our daily activities, that is, their structural properties, arise from the interaction of multiple aspects of a setting including psychological, material, social, historical, political, and economic factors as these are seen by the actors themselves.

Science and Activism

Given that only a small fraction of students eventually become scientists or engineers, one has to question traditional models of delivery which make science education a propaedeutic (and sorting) effort. This begs the question, "What other reasons could there be for science education?" Some answers may come from science and technology studies. Research in this domain shows that people from all walks of life, even without scientific or technological training, can have an enormous shaping influence on the nature of scientific inquiry. This includes the determination of validity, reliability, and appropriate testing protocols (e.g., Blume, 1997; Epstein, 1995; Solomon & Hackett, 1996; Rabeharisoa & Callon, 1999). A survey of 430 recombinant DNA scientists engaged in research showed that only 6% thought that public attention has not had an impact on their work (Rabino, 1991). Forty-four percent of the scientists thought that the impact has been beneficial while the remainder suggested that the impact has been harmful (24%) or equally beneficial and harmful (27%).

This research also shows that people who identify with some genuine and authentic (because it is relevant to their lives) cause—activists for various causes such as the environment, AIDS, medical treatments—engage science such as to change the very nature of what we understand science to be. That is, the nature of scientific experiments, who is to participate as subject, how controls are to be instituted, and so on is negotiated between scientists and activists, often in the forum of legal courts. That is, without scientific training (indoctrination?),[5] these groups engage science in a socio-political process and thereby shape what science is and how it is done (and, as we suggest in our other chapter, also change the political process). Individuals and groups who are not "card-carrying" scientists increasingly influence science, scientific validity, and approaches to science. These non-cardholders include judges, activists, consumers, engineers, and a whole host of others engaged across a broad range of sites including legislatures, courts of law, sites of activism, the press, and the marketplace (e.g., Clarke & Montini, 1993; Solomon & Hackett, 1996).

---

5. Becoming a scientist means that we are also becoming members of a scientific discipline. But becoming a member of a discipline involves being disciplined, submitting to a discipline, often in a very physical sense of the word (Roth, 2001; Roth & Bowen, 1999b).

The best-known case study about the changing face of science through the participation of activists comes from AIDS/HIV. AIDS activists, with little formal training in science or medicine, have constructed themselves as credible actors and genuine participants in the design, conduct, and interpretation of clinical trials used to test the safety and efficacy of AIDS drugs (Epstein, 1995, 1997). Because the activist groups were recognized as the legitimate voice of groups of afflicted individuals, they were able to position themselves as obligatory passage points for researchers seeking participants in clinical trials. Activists also learned to draw on scientific registers and discourse patterns when they participated in public discourses at scientific conferences, perused research protocols, or consulted with sympathetic professionals (see also our case study in the Lee and Roth chapter). Once the activists' arguments drew on the registers of medical science and biostatistics, researchers were forced, by the norms of their own discourse and behavior, to consider the arguments of the activists. Activists used their newly acquired registers and discourse competencies to construct powerful arguments (as one expects from Latour's [1987] Machiavellian actors). Activists emphasized the right of individuals to assume the risks inherent in testing protocols, and they insisted that the individuals involved in the clinical trials should be representative of the broad spectrum of groups affected by the disease. Through their participation, activists shifted the focus of clinical trials from scientifically relevant questions to the construction of information useful to patients and doctors in their ongoing clinical praxis. These changes produced a shift from purely scientific interests to one in which the needs and interests of those who were actually afflicted with the life-threatening illnesses were addressed.[6]

Because of their success, AIDS activists have had a secondary influence on science in as far as they served as a "role model" for other health-related activists. For example, the support groups for Lyme disease patients successfully deployed their arguments be-

6. Some readers may be interested in the extended case study of AIDS treatment activists' influences on beliefs about the efficacy of a combined drug therapy (AZT and ddC) (Epstein, 1997).

fore the organizers of an international conference on Lyme disease in order to reinstate conference papers that had initially been rejected. Chronic fatigue patients launched a successful lawsuit against a drug manufacturing company that violated its promise to continue to supply an experimental drug after the clinical trial ended.

Community, School, and Environment

Our on-going research is situated in a local community where environmental activists, farmers, local residents, and school children (middle, high school) each take an interest in one of the watersheds in which their community is located (Lee & Roth, this book; Lee & Roth, 2001; Lee, Roth, & Bowen, 1999). Our research provides detailed descriptions of how people enact science education in and out of schools. They learn science *not* for its own sake, but because they are interested in the goal of making the community a better (ecological) place to live, to make it into a "dwelling," and to preserve (and restore) the watershed so that it can survive as an ecological whole. Here we want to provide a few images of a curriculum project designed to have middle school students participate in, and contribute knowledge to, the community in which they live. In this way, science is no longer boxed but takes on significance as part of a larger community.

Images from the Oceanside Middle School Project

Our curriculum unit on environment and ecology begins with an article from the local newspaper which the students in this grade 7 read and discuss in class taught by Michael Roth and Mrs. Roberts (pseudonym), the regular science (and mathematics) teacher of the class. The article emphasizes, among others, the following points that become aspects of the subsequent whole-class discussion:

> Beaches are closed due to high coliform counts. That can be fixed, but it's going to cost. The question is: Where do you want to live?

The wider community must be involved in the physical tasks and...
the wider societal and political issues surrounding the problems. (Reimche, 1998, p. 9)

In the whole-class discussion, students consider it as an important
and worthwhile task to contribute to the community effort by doing research along Henderson Creek. Among others, some students
recognize that the (fecal) bacteria and chemical load on the creek
ultimately becomes a load on the nearby inlet including fish and
shellfish, and ultimately on the local fishery and the families that
draw their livelihood from it. The students immediately suggest a
range of actions, such as cleaning up the creek, but the teachers
invite them to begin with learning more about the creek, the surrounding areas, the ecology of the creek and surrounding areas,
etc. Before ending this introductory lesson, the teachers suggest
that students should think about presenting the knowledge they
gain back to the community. In addition to students' suggestions
to report to their peers (in the school) and to their parents, the
teachers make students aware of the open-house event arranged
by a local activist group, which will occur four months later in the
year. (For a description of some of the activists' practices, see Lee
and Roth, this book.) The teachers also suggest that students
might want to think about producing a web page where they report the results of their research.

During the following lesson in the field (Mrs. Roberts had con-
tacted parents and five of them had agreed to serve as drivers),
the teachers and two graduate students (Stuart, Michael Bowen)
each take a group of students to different sites along the creek.
One of these sites is the riffle constructed by the local activist
group (see Lee & Roth, this volume). In this way, students are
provided with opportunities to construct a bigger picture about
the placement and role of the creek in the community and environment. Students begin by framing research questions they could
investigate during the following lesson at different locations along
the creek. One group decides that they want to test for the presence of coliform bacteria at different sites, another group is interested in finding out more about the organisms that live in different
parts of the stream and along its banks. A third group is inter-

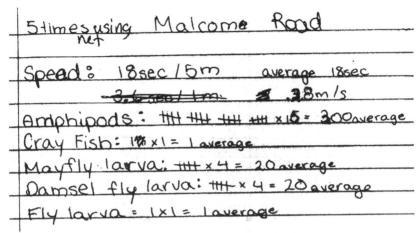

Figure 1. Records kept by one group of students as their contribution to the class investigation of the relationship between rate of flow and frequency of several invertebrates.

ested in the types of plants that border the creek and populate the surrounding areas.

The activities of *these* students are connected to those of other students as well. Thus, students from another class who have already completed a unit of ecology with the concomitant research along the creek initially accompany Mrs. Robert's students to serve as peer consultants. Sometimes, Meagan, the local environmental activist, and Karen, a water technician working for a local farm, also come to observe, interact with students, or provide a demonstration. Sometimes, a parent is also interested in working with a group of students to scaffold their data collection efforts.

Later in the unit, the teachers communicate to students the concern that there do not seem to be sufficient data to make claims that can be substantiated during the public demonstration. As a class, students decide to conduct measurements that would answer one common question, "Is there a relationship between the type and frequency of organisms living in the creek and the speed of the water?" During the following double period which the class spends in the field students use part of their time sampling different sites for invertebrates (using D-nets and Serber samplers) and measure stream speed at their location in one of two sites where the research is conducted. As always, there are interested parents who not only drive students to the different sites along the creek, but who actively work with groups of children to scaffold their data collection effort. Here, students learn to use tools and instruments in such a way that they can make claims that are not easily deconstructed by others.

First, students collect invertebrate samples from the stream bottom using a Serber sampler (which has an area of one square foot). To do this they place the grid on the bottom of the creek with the collecting net flowing downstream and, using their hands, stir up the creek bottom within the grid breaking up clumps of mud and lightly rubbing the surface of any rocks or debris. Carefully rinsing the sample in the net (to remove mud), they transfer the sample to a small plastic container and mark its collection site. Then, using floating wooden blocks, students collect data to determine the creek speed at the sites at which they collected organisms in previous weeks. In the team, one student drops a wooden block, another keeps track of time using a stopwatch, and a third calls out when the wood arrives at the five-meter mark. Each team replicates the speed trials at least three times and then calculates an average. The data collected in the field by one group, and as recorded in their field notebook, are presented in Figure 1.

On the following day, students do initial analyses and begin mathematization and thereby prepare subsequent mathematical analyses. They analyze their water samples, separating the organisms and using a microscope and pictorial keys to identify them. First, students pour the water from their buckets into trays, which they then inspect for organisms. Whenever they find one, they take a turkey baster and, by releasing the pressed bulb, draw it up. They then release the captured creature into a depression of an ice cube tray. When students are done, they count the number of individuals from each species in their own samples and record their results in a data table drawn on a transparency. They also make sure to appropriately categorize all the creatures using a chart as a referent during their microscope work.

During the subsequent science lesson, students make further progress in the construction of claims about the creek ecology that

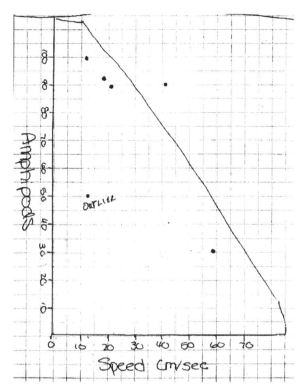

Figure 2. Graph produced by students to find out if there is a relationship between the speed of Henderson Creek and amphipods.

are supported by data. Each group examines the entire table for patterns in the class data from which they produced graphical inscriptions. One group produces a graph depicted in Figure 2. In attempting to make sense of the data, students note that one point seems to fall outside a simple relationship. One of the graduate students (Mike B.) present on that day talks to them about "outliers," a concept which they could use to eliminate from consideration certain points (if small in number) from considerations. The students in that group conclude that there exists a relationship between the numbers of amphipods and the creek speed. Elaborated in their conversation is the possibility that, even though there was probably more oxygen where the stream speed was faster (this had been discussed earlier) something else was affecting the presence of amphipods. This might include being washed away by the higher speeds or perhaps the presence of a different predator.

Towards the end of the unit, two groups of 7 students each produce a class poster. The students decide that before they present their work in public, they want to conduct a trial presentation in another grade 7 who will soon begin to conduct their own research in the watershed. The following excerpt from our video records shows a brief exchange between our students (Niels, Ella) and the teacher and a student from the other class (Mrs. Nelson, Lisa) in which students display what they have learned from their studies. The excerpts show that even normally low-achieving students such as Niels develop discursive competence drawing on the resources of ecology.

Nelson:	We are probably seeing different things, and probably at different stages than you guys did?
Niels:	You guys probably will see other things like crayfish and other fish. Perhaps not other fish, but crayfish.
Lisa:	Why do you think so?
Niels:	I am not sure, but probably because it is warmer. We found, when we were going it was winter, we found nothing, no fish, nothing. But as soon as the temperature got warmer, we started finding more mayflies . . .
Ella:	More crayfish, larva, dragon flies . . .
Niels:	And arthropods. I found like four of those in just one scoop.
Lisa:	What other kinds of things did you find?

Here, students not only point out the types of creatures they observed but also the differences in the amount during winter and spring months.

The final stage and ultimate purpose of the unit is the open-house event organized by the members of the Henderson Creek Project: Here students contribute the knowledge they constructed to the community at large and have opportunities to engage with visitors. The two groups constructed posters where they, among others, report the relationship between stream speed and the various organisms (amphipods, damsel fly larva, fly larva, and mayfly larva) that they had identified and counted. There is also a scale model of one of the research sites, and the equipment used to identify the organisms (ice tray for sorting, turkey baster for collecting organism from tray, trays, 2 microscopes). The students also have set up a computer on which they present the web site that they had constructed including the photos they had shot using our electronic camera. Our video shows people milling about, interacting with the web site, studying students' equipment, and asking students questions about the research they completed. Though Jimmy is not a high achiever in his normal classes, he competently explains to a visitor of the Open House what they had done.

> Visitor: Where did you get the creatures?
> Jimmy: We got them at Miller Road. That is by, what's that school
>   called, it is by Valley Elementary School. You take the path,
>   all the way down and you get to the creek. We went there and
>   we got so many bugs in just five minutes. And then we
>   counted, what our whole class gets, and that is about 30 of
>   us, and we got in five minutes that much.

In the end, the members of the environmental activists suggest to us that the Open House was a great success. Of course, they note, there is an interaction between their own efforts and the fact that the students present their work. That is, the children's work brought many parents to the Open House and therefore increased the environmental activists' exposure to the members of the community. In this way, the school children contribute to the extension of the network of people being and becoming aware of the environmental issues of the watershed in their community and the impact their own life styles have on this aspect of the environment.

Reflections on the Project

Here we briefly reflect on the significance of children's activities. The activities we described are more than just school activities. These activities are connected into a larger set of activities that give rise to learning in a community larger (and with more learning opportunities) than the classroom communities that we had conceived of in our earlier work (Roth, 1998a). As with other experimental units, we find that students often marginalized in regular class science lessons (such as Niels and Jimmy) "find themselves" in these activities that allow them to set their learning agenda, by asking questions that they answer themselves. Throughout the unit, the teachers and students reflect repeatedly on the fact that the unit is designed for students to contribute to the community knowledge about the watershed. As part of the process of contributing to the overall project, students learn to sample, use scientific tools, and engage in mathematical analysis (e.g., comparing rates of flow when different distance measurements were at the base of their timing). They also learn to engage in rhetoric using claims and evidence and develop a sense for the aesthetic of place in the community. Most importantly, the activities are connected to the activities of others and thereby contribute to the network of activities that stabilize environmental concerns in the community. Students from other classes who have already more experience contribute to the learning of new classes in the project. Parents participate not only in driving children to the different sites, but, after being introduced to the pedagogy motivating the unit, also work with small groups of students, and activists and scientists also participate in the activities of the school. Finally, the very fact that students do their research out in the field, at least part of their school day is spent *in the community*. By contributing to the knowledge base of this community at large rather than studying for the sake of studying in a building officially designed for such activities.

The students also look forward to doing their fieldwork. Thus, after the two days of researching the relationship between organism type and frequency and stream speed, Anne wrote:

> I learned that different kinds of bugs like to live in different areas of the creek. I also learned how to measure the speed and acid level of the creek. I think that what surprised me most was the number of amphipods that were in the area of the creek my group and I studied. I liked using the Serber net the best, and then emptying it into a container. I also liked looking at the bugs through the microscope. The worst was that I didn't have rubber gloves and the water was cold and dirty. We worked together OK, except for John flinging water at Sandra.

Interactions with teachers and peers are consultative rather than top down. The different nature of the curriculum was also evidenced by other activities surrounding the work in which the students were engaged. Parents contributed time by participating in driving students to the study sites, and many eagerly participated by asking students questions, scaffolding methodological approaches, and helping the students interpret their findings. The regional mayor visited the students as they worked at their field sites and questioned them about their work. In another aspect of their work, when they were working on a report to present at the open house of the Henderson Creek Project, the mayor was available to be interviewed and discussed the creek with them and what contributions various groups, including theirs, were making to improving the watershed. Field scientists from this work participated with the students in the field and came and worked with them in their classroom as they worked at making sense of the field samples they had collected. Karen, the water technician from a local farm, also assisted students to collect biotic and abiotic information at various sites along the creek. Parents, including those from the local First Nations communities accompanied students, teachers, and scientists.

In this description, we see how members of different parts of the community—parents, environmental activists, scientists, teachers, preservice teachers, students—move in and out of their respective sub-communities. Each time there is some contact, we see learning events where knowledge from one group in the community is made available as a resource to another group. Thus, even if a member of one group interacts with another and therefore, in some sense, is a legitimate but peripheral participant, it is possible to document learning. For example, Michael Roth also worked with a teacher intern implementing an environmental unit. She learned tremendously about science, teaching, and learning in communities as she interacted with him, an experienced science teacher and scientists (e.g., Roth & Boyd, 1999; Roth, Masciotra, & Boyd, 1999). The Henderson Creek Project scientists and environmental activists also learn from the data collected by the students and about the impact science education can have on the community; they learn about the challenges of teaching and of getting young people interested in the issues that will concern them tomorrow. Parents and other members of the community learn about Henderson Creek from the children, and the First Nations children (normally attending their own tribal school) learn from the experiences of the Oceanside Middle School students through Michael Roth's teaching.

Science Education for Citizenship

Places for Science Education: Fields of Social Action

Toady's world is becoming increasingly complex. More and more, the knowing of past generations is embodied in the technologies we use. Global communication patterns interact with our traditional ways of getting on in our work and lives, sometimes stabilizing, sometimes destabilizing what we used to do. In such a world, being aware and participating in the negotiation of issues that necessitate particular knowledge of the environment become increasingly important. Developing scientific literacy means that as individuals, we are enabled to increasingly participate, in informed ways, in some issue critical to our future. What we do not need is decontextualized and memorized meaningless bits of information, but an active knowing how to participate in the various ongoing conversations, each drawing on particular or diverse discourses for playing out a variety of interests. What we do not need is a science education that mainly acts as a filter to shake most people out of science, and only retain those few who will study it as a part of their future scientific careers (only about 5% of the student population that enters high school). In almost every community, there are many issues that students can investigate and, as they take action, can contribute to the conversation at large. We already pointed out what we have done with the students and teachers in our own community. But there are many other issues, which students could research, contribute to, become involved in, etc. Consider the following examples from different areas in the world where science and business have contributed to havoc in natural animal populations. But other examples where the interaction of science and business have had a positive impact could become starting points for students' inquiries such as plant nurseries, socially active businesses, etc.

Traditional Ecological Knowledge and Human Actions

There is an interesting tension between traditional ecological knowledge (and by this we mean more than the knowledge of First Nations People, but also that knowledge which was developed in the tradition of Western, non-science practitioners) and scientific knowledge. In our minds, it is quite clear that the approaches of formal science to knowledge construction are different from other approaches. That is why traditional ecological knowledge is not science, but it constitutes a body of accumulated (discursive and material) practices that has its own value. However, in today's world, these two forms of knowledge are often played out against each other to the detriment of nature. Here, in this debate between the different forms of knowing, there are places for participating in and contributing to the debate. Take the following three sketches as examples of such potential places for engaging in relevant, purposeful, and concernful action. (For more, detailed examples see Roth, in press-a, in press-b.)

> Story 1. Hong Kong businessmen offered the Solomon Island chiefs $5 Sol. Isl. currency per kilo of live reef fish caught. The fish sold for $300/kg in Hong Kong. The businessmen encouraged the Solomon Islanders to fish the reef during spawning season for the fish, when they would aggregate in big spawning tangles. Previously they would fish these tangles very lightly. The Islanders, against all tradition and traditional ecological knowledge, began to fish heavily during spawning season. Lo and behold, the fish stocks crashed in about four years.

> Story 2. In Newfoundland and Labrador, native people (Inuit, Montagnais), locals of European descent, and European fishing fleets (Viking, Portuguese, Basque, Spanish) have fished cod for centuries. In more recent years, fisheries officers determined fishing quotas based on scientific models. About 10 years ago, the cod population crashed wiping out virtually all schools that had come near the shore. The locals claim that the quota system (and large off-shore fleets) had brought the cod to extinction.

Story 3. On the Olympic peninsula in the Pacific Northwest, the young members of a tribe are planning to re-institute whale hunting, which the tribal forefathers had given up some 70 years ago. They already have sold their planned catch to Japanese merchants. Activists from the area and around the country mount protests and vow to prevent the First Nations boat(s) from getting to their targets.

In all three stories, traditional practices and traditional knowing are in conflict with political, economic, or environmental interests. Being scientifically literate, from our perspective, means that we are enabled (increasingly so) to participate in the conversations around the issues associated with fishing in the three places. There are many other ways fish, fish stocks, fishing industry, sports fishing, fish habitat restoration, natural and life history of different fish, and so forth could become central to the activities in classrooms. Furthermore, fish are but one of an almost infinite range of topics that can be investigated by school children and older students, or that entire classrooms of students can choose as focus of social, political, and ethical actions.

We do not pretend to have an answer how to approach a particular issue independent of its context; even less do we want to pretend to have answers to the current malaise in science education. We can no longer pretend that science and scientific knowledge are apart from society, are inherently value-free or value-neutral and apolitical (see also Lee & Roth, this book). Rather, science and scientific knowledge are tied up in complex actor networks that continuously shift and adapt to local pressures and activities. New nodes develop; old nodes are eliminated or collapsed into a black box, which subsequently acts as a new, possible more powerful actor in a redefined network and nodes that redefine themselves. Scientific literacy means to be able to participate in this continuous weaving and reweaving of society as it concerns science and scientific issues, particularly environmental issues. Science education can therefore no longer be a filtering and job training mechanism for the few who survive and then become scientists. Rather, scientific literacy has to enable individuals to participate in these ongoing discourses, has to enable us to become activists and engage with scientists whatever the cause (environment, AIDS, genetic engineering of food, legalization of drugs, etc.).

Concluding Thoughts

In this chapter, we argue that if we want to do "authentic" school science we have to reconsider what the science is that students are to participate in. Science can no longer mean doing and encul-turating students into the science of scientists as we have con-ceived of it in our own earlier work (Roth, 1995). If we attempt to do this, we continue to deprive many of their right to become sci-entifically literate. In contrast, we argued that being scientifically literate means being enabled to participate in discourse, and ac-tion related to issues where scientific knowledge (that is, scientific discourse) comes in handy as a resource and referent. Knowing means competence to engage in relevant action; learning therefore requires participating in real, everyday science from early on be-cause relevance, purpose, and concern are tied to these types of activities. In contrast to previous conceptions of science educa-tion, this participation can begin early when students engage in actions that have impact on their everyday life, their schools, communities and families, whatever the students' age. In this way, science is no longer an activity isolated between the walls of schools and something with a purpose years down the road. Rather, "authentic" school science means participating today in making this a better world. "Authentic" school science is then no longer *just* job preparation but a participation (with increasing levels thereof and therefore increasing competence) in everyday life affairs through activism, stewardship, or simply as a con-cerned citizen. "Authentic" school science and scientific literacy therefore are forms of social actions and knowledgeability for so-cial action and for accessing science. Science education, as distinct from science schooling, means to participate—and to do so in in-creasing ways—in everyday issues that may require science as a referent and resource, as a matter of course, and in a purposeful way. Given that few students actually become scientists, we see in "authentic practices" to be fostered in schools those of lay scientific pursuits—such as those in nature clubs, environmental activist groups, or just plain concerned citizens tending their gardens—not just the practices in scientific laboratories. Such a reconceptualization changes school science from a propaedeutic effort that aims at training scientists to a continuing participation in everyday science-related activities. In this new view, continuous trajectories of learning and membership compatible with a strong view of situated cognition are possible and feasible. Such changes would entail a deinstitutionalization of school science where activities are evaluated in terms of their contribution to a common good (whatever the community negotiates this to mean) rather than in terms of individual memory and intellectual prowess for processing information.

We propose that science education ought not to content itself with teaching students science, or even how to construct evidence so that it becomes more unassailable and therefore better supports the claims they make. (Activists enact the same or similar practices as we show in Lee and Roth.) Rather, science education should enable students to engage in the community for a larger good, as this is defined by the group. Science needs to be regarded no longer as something apart, practiced by a few high priests, and in special places.

Such changes should occur with a concomitant change in curricula that allows students to enact and develop their epistemological discourse practices and take a more reflective stance to all forms of knowledge, scientific or otherwise (e.g., Désautels & Roth, 1999). Interrogating knowledge and knowledge construction, of course, is at the heart of border pedagogy, itself concerned with educating students for life in an open and free society (e.g., Giroux, 1992). Although we remain science teachers and science educators at heart, we are also concerned with the larger issues of *cui bono?* We therefore have to ask what it is about science that can help us to lead away from the reproduction of inequalities along the lines of gender, race, social class, and expertise (cf. how "scientific" testimony outweighs other testimony in courts, construction of "learning disabilities," etc.). We side with calls for an ideal democratic (and therefore free) community in which reigns a sense of solidarity (Rorty, 1991). In such a community, absolutist discourses (scientific, religious, or otherwise) do not have a place. Rather, all forms of discourse are open to interrogation, subject to a social epistemological inquiry of recontextualizing that no longer requires "science" and "scientific" for taxonomic purposes.

Stuart: I think that activism is a great way to learn about science. It is the only site in science where you get everything—people do science; people critique it—you get access to all kinds of science—industry, academic, government, other values are taken seriously like aesthetics; science is seen and talked about as related to business and politics and history. It is the confluence of many different streams of discourse. Through my experience with activism, I learned painfully about the political agenda that science can carry. I learned how science couldn't really save the world, but that its value is ultimately measured by society. And I, a typical isolated lab researcher, got an appreciation for science's multiple connections with the rest of society. I also dreamed up this vision of the wonderful good that could come out of a citizenry of people. These people are not intimidated by scientific jargon or the tone of experts but are able to deconstruct authorities' arguments and take action to make changes that benefit their lives and communities. This vision still calls me.

Michael: For me, engagement in the concerns of an increasingly smaller world—small because we can get anywhere in a short amount of time, and because of an increasing population density giving each of us less space—should become part of our lifeworlds. There are activities around the house that are part of a different scientific literacy. Such activities include composting, reducing the amount of garbage that goes into land fills, and producing fruits, vegetables, and herbs to reduce the dependency on food shipped in at high cost from elsewhere. These activities, once enacted as part of everyday life, contribute to make this a better place to live, a dwelling.

Acknowledgments

This research was supported by grants 410-96-0681 and 410-99-0021 from the Social Sciences and Humanities Research Council of Canada. The views represented in the chapter are our own.

References

Blume, S. S. (1997). The rhetoric and counter-rhetoric of a "bionic" technology. *Science, Technology, & Human Values, 22*, 31–56.

Bourdieu, P. (1997). *Méditations pascaliennes*. Paris: Seuil.

Brookhart Costa, V. (1993). School science as a rite of passage: A new frame for familiar problems. *Journal of Research in Science Teaching, 30*, 649–668.

Brown, J. S., Collins, A., & Duguid, P. (1989). Situated cognition and the culture of learning. *Educational Researcher, 18*(1), 32–42.

Clarke, A., & Montini, T. (1993). The many faces of RU486: Tales of situated knowledges and technological contestations. *Science, Technology, & Human Values, 18*, 42–78.

Derrida, J. (1988). *Limited inc.* Chicago: University of Chicago Press.

Désautels, J., & Roth, W.-M. (1999). Demystifying epistemology. *Cybernetics & Human Knowing, 6*(1), 33–45.

Dewey, J. (1933). *How we think*. Boston: Heath.

Epstein, S. (1995). The construction of lay expertise: AIDS activism and the forging of credibility in the reform of clinical trials. *Science, Technology, & Human Values, 20*, 408–437.

Epstein, S. (1997). Activism, drug regulation, and the politics of therapeutic evaluation in the AIDS era: A case study of ddC and the 'surrogate markers' debate. *Social Studies of Science, 27*, 691–726.

Fourez, G. (1997). Scientific and technological literacy as a social practice. *Social Studies of Science, 27*, 903–936.

Fuller, S. (1997). *Science*. Buckingham: Open University Press.

Giroux, H. (1992). *Border crossings: Cultural workers and the politics of education*. New York: Routledge.

Hazen, R. M., & Trefil, J. (1991). *Science matters: Achieving scientific literacy*. New York: Doubleday.

Heidegger, M. (1977). *Sein und Zeit*. Tübingen, Germany: Max Niemeyer.

Hutchins, E. (1995). *Cognition in the wild*. Cambridge, MA: MIT Press.

Latour, B. (1987). *Science in action: How to follow scientists and engineers through society*. Milton Keynes: Open University Press.

Lave, J. (1988). *Cognition in practice: Mind, mathematics and culture in everyday life*. Cambridge: Cambridge University Press.

Lee, S., & Roth, W.-M. (2001). How ditch and drain become a healthy creek: Representations, translations and agency during the re/design of a watershed. *Social Studies of Science, 31*, 315–356.

Lee, S. H., Roth, W.-M., & Bowen, G. M. (1999, March). *Learning at the interface: Knowledge transfer and transformation between heterogeneous and overlapping knowledge-building communities*. Paper presented at the annual conference of the National Association for Research in Science Teaching, Boston, Mass.

Maxwell, N. (1992). What kind of inquiry can best help us create a good world? *Science, Technology, & Human Values, 17*, 205–227.

McGinn, M. K., & Roth, W.-M. (1999). Towards a new science education: Implications of recent research in science and technology studies. *Educational Researcher, 28*(3), 14–24.

Orr, J. (1998). Images of work. *Science, Technology, & Human Values, 23*, 439–455.

Rabeharisoa, V., & Callon, M. (1999). *Le pouvoir des malades*. Paris: Écoles de Mines.

Rabino, I. (1991). The impact of activist pressures on recombinant DNA research. *Science, Technology, & Human Values, 16*, 70–89.

Reimche, J. (1998, December 16). Group is a bridge over troubled waters. *Peninsula News Review*, p. 9.

Rorty, R. (1991). *Objectivity, relativism, and truth: Philosophical papers* (Vol. 1). Cambridge: Cambridge University Press.

Roth, W.-M. (1995). *Authentic school science: Knowing and learning in open-inquiry laboratories*. Dordrecht, Netherlands: Kluwer Academic Publishing.

Roth, W.-M. (1998a). *Designing communities*. Dordrecht, Netherlands: Kluwer Academic Publishing.

Roth, W.-M. (1998b, November). *Job preparation or social activism and stewardship? 'Authentic' school science at the cross roads*. Paper presented at *Making Connections*, Baltimore, MD.

Roth, W.-M. (2001). 'Authentic science': Enculturation into the conceptual blind spots of a discipline. *British Educational Research Journal, 27*, 5–27.

Roth, W.-M. (in press-a). Taking science education beyond schooling. *Canadian Journal of Science, Mathematics, and Technology Education*.

Roth, W.-M. (in press-b). Science in/for the community. *Enseñanza de las Ciencias*.

Roth, W.-M., & Bowen, G. M. (1999a). Digitizing lizards or the topology of vision in ecological fieldwork. *Social Studies of Science, 29*, 627–654.

Roth, W.-M., & Bowen, G. M. (in press). Of disciplined minds and disciplined bodies. *Qualitative Sociology*.

Roth, W.-M., & Boyd, N. (1999). Coteaching, as colearning, in practice. *Research in Science Education, 29*, 51–67.

Roth, W.-M., Masciotra, D., & Bowen, G. M. (in press). From thing to sign and 'natural object': Toward a genetic phenomenology of graph interpretation. *Science, Technology, & Human Values*.

Roth, W.-M., Masciotra, D., & Boyd, N. (1999). Becoming-in-the-classroom: a case study of teacher development through coteaching. *Teaching and Teacher Education, 17*, 771–784.

Roth, W.-M., & McGinn, M. K. (1997a). Deinstitutionalizing school science: Implications of a strong view of situated cognition. *Research in Science Education, 27*, 497–513.

Roth, W.-M., & McGinn, M. K. (1997b). Science in schools and everywhere else: what science educators should know about science and technology studies. *Studies in Science Education, 29*, 1–44.

Roth, W.-M., & McGinn, M. K. (1998). >unDELETE science education: /lives/work/voices. *Journal of Research in Science
Teaching, 35,* 399–421.

Roth, W.-M., McGinn, M. K., & Bowen, G. M. (1996). Applications of science and technology studies: Effecting change in
science education. *Science, Technology, & Human Values, 21,*
454–484.

Roth, W.-M., & Milkent, M. M. (1991). Factors in the development
of proportional reasoning by concrete operational college students. *Journal of Research in Science Teaching, 28,* 553–566.

Saxe, G. B. (1991). *Culture and cognitive development: Studies in
mathematical understanding.* Hillsdale, NJ: Lawrence Erlbaum
Associates.

Scribner, S. (1986). Thinking in action: some characteristics of
practical thought. In R. J. Sternberg & R. K. Wagner (Eds.),
*Practical intelligence: Nature and origins of competence in the eve-
ryday world* (pp. 13–30). Cambridge: Cambridge University
Press.

Solomon, S. M., & Hackett, E. J. (1996). Setting boundaries between science and law: Lessons from Daubert v. Merrell Dow
Pharmaceuticals, Inc. *Science, Technology, & Human Values, 21,*
131–156.

CLUSTER II

From Teaching STS to Enacting Science at the Grassroots Level: Shifting Agency

Conversation 2

Michael: This second set of chapters illustrates our commitment to a form of epistemological democracy whereby different positions are not set one against the other, but rather they are brought together in a dialogue. It begins with the piece by Cross and Price, who argue strongly for what they call a socially responsible science and science education. They frame their argument within the STS orientation and highlight a series of controversial topics such as food manufacturing that could be the focus of modern curricula redesigned to socialize current issues. But the authors, probably in a realistic way of thinking about what can be done here and now in a majority of schools, claim that it is not necessary to do away with the conceptual structure of the customary science education curriculum.

Jacques: From what I know about science education in our schools what they propose amounts to a radical change in the pedagogical habitus. The only time the students hear the word "social" in science classrooms is when a teacher says that technology, the application of scientific knowledge, can have social effects.

Michael: I understand what you are saying, but one can argue that what teachers consider to be an important issue may not be a source of mobilization for the students. At least, granted that one case study cannot constitute a proof, we see how Tobin's students resisted his teaching, even though he made all effort possible to select topics *he* thought could be a source of meaning for the students.

Jacques: You have a point here, but it would be quite possible to have the students participate in the definition of what constitutes worthwhile problems to be explored. The teacher could then help students make connections with the controversial topics alluded to by Cross and Price. In fact, I think that their position is somehow quite close to Tobin's as far as scientific knowledge is concerned. They all seem to claim that we should not throw away the baby with the bath water and that we should keep on teaching some classical science concepts.

Michael: I would like to remind you what our elder said at the end of his chapter, namely that the idea of science education for social action had him think that we should ask ourselves the following question. What is school science for? And maybe Aikenhead, Barton and Osborne and Lawrence and Eisenhart provide answers to this question, which differ from the previous positions. At least their insistence on the involvement of the community in the definition of what is considered proper knowledge can be interpreted in this way.

Jacques: In other words, you are saying that these authors are actively relinquishing their grip on the curriculum, which provides a space for either First Nations people or people in other communities to frame what the relevant knowledge might be that we should teach children.

Michael: There are other epistemological and ideological stakes involved, but I hope that our presentation of this cluster will be seen as an invitation for readers to participate in the dialogue initiated by the authors.

5

Teaching Controversial Science
for Social Responsibility:
The Case of Food Production

Roger Cross & Ronald Price

While the science we discuss here is more suitable for older students, those in the last three years of schooling (grades 10–12), science teaching consistent with social responsibility needs to start when science is first taught and then followed through to the senior years. This will require changes of teaching style to emphasize the produced nature of scientific theory and the functions that different kinds of theories serve. Our concern is a science for everyone. But at the same time we are mindful of the needs of the minority of students who will go on to study science further and even become scientists and technologists in later life. In contrast to Ziman (1980), we argue that the kind of approach to science teaching we are advocating here is at the same time a good and sufficient, if not the best, basis for more advanced studies. There is an urgent need for a new generation of scientists and technologists who have a better understanding of the nature of science, who can understand it in its social context. If they are not trained differently the much-needed reform of science may not occur. In addition to the concepts and skills of traditional science courses, the kind of teaching we are advocating will introduce that interest in, and concern for, relating science to the needs of the wider community which is so essential today for specialist and lay person alike.

One of the objections which has been advanced against teaching controversial topics, whether as part of integrated science or some form of social responsibility orientated topic approach, is that it is not "systematic." That is, it fails to teach "essential" knowledge and skills or that it teaches them in a less satisfactory way than traditional subject (discipline) teaching (e.g. force, energy, in physics). Of course there are some things which must be taught before others, some concepts which depend on others for understanding and any teaching which ignores this can only land the teacher and students in trouble. But this difficulty is not avoided by keeping to subject matter teaching as any careful examination of syllabuses will reveal. Nor is it simply a question of the syllabus. Teaching and learning are involved in integrated science, and learning does not easily keep pace with the most ideal syllabus, especially when one is thinking of a class of 20-30 students with all the learning dissimilarities that implies. If it is a question of alleged neglect of essential knowledge and skills, a comparison of the "fundamental knowledge," the "basic concepts" that form the backbone of courses taught around the world, and the knowledge required by a socially responsible approach to science would reveal no such neglect.

Many will fear that by focusing on social issues and social responsibility in science we shall replace the serious teaching of scientific principles with a superficial and wrongly biased account of what is the proper concern of the social sciences (economics, sociology) or history. We hasten to admit that there is such a danger, and that in some cases this might occur. But the most rigorously discipline-oriented syllabus also cannot prevent bad teaching. One difficulty is that cooperation between subject teaching is so hard to achieve in the normal school situation. A certain amount of discussion of the social setting of scientific questions has always been necessary. The approach we advocate will require more and will also require cross-subject cooperation for maximum success. But whatever the particular school situation, our concern is the special responsibility of the science teacher as a consequence of his or her special training. This is the teaching of the science that will allow judgement of the social issue to be optimally based.

There remains the problem of sequencing material so that it is introduced at a suitable level of difficulty and that there is not boring repetition in different grades. There is the problem that much of the science involved in social issues would seem at first glance to be much too difficult for treatment even in grades 11–12, much less 7–10. One can only reply that (a) something must be done to help future concerned citizens and (b) science teachers are already familiar with the problem of making difficult material understandable at different levels of simplification. What is needed is much more discussion and examples of how to do this with the essential science for social responsibility so that the simplification does not distort. Another problem which may be worrying is that repetition of topics, e.g., models associated with nuclear energy, may reduce interest when they need to be presented again at greater depth in the higher grades. This problem is certainly a real one and one that has plagued science teaching for a considerable time. There are spectacular demonstrations being repeated ad nauseam. But the objection can be met by experimenting with a spiral curriculum, which demonstrates the value of repetition with increasing complexity and perhaps wider application.

The Role of Experts

A common thread running through all scientific and technological controversies is the role of experts. The controversies surrounding food give us many lessons about the nature of science, including: the sociopolitical interests of scientists, social responsibility (control of knowledge, who can and cannot speak with authority in science, suppression of debate, the networks that protect careers), and the power of experts.

The influence of experts in today's society—here scientific experts—poses particular threats to a democratic society (Martin, 1996). This is clearly seen in the many problematic areas of science, particularly where the security of human health and food is concerned. The problem of the expert is at the center of the difficulty of the democratic participation of the public in debate and decision-making. It is the role experts are given and the general contempt with which ordinary people are held when attempting to participate that makes it so difficult for the public to have a say. As citizens we need to assiduously probe the interests which may be underpinning the opinions of experts, for as Joseph Rotblat said, "when expert scientific judgements are mixed with political, social and economic decision-making, controversy is the rule" (1981, p. 32). Judge Bazelon (1979) refers this mix of values and science as being pervasive in expert opinion and the blurred boundaries as trans-science.

The biggest inhibitors confronting the public, and a more mature relationship with science, is the assumed authority of science to speak "truth" statements, this tends to make the public impotent when confronted with scientific experts. This is actively promoted by the institution of science and involves the kind and status of the knowledge that is so often claimed to be infallible and impersonal. This situation is promoted wittingly or unwittingly through the schooling of science. Another problem is the language of science, which creates a discourse barrier—those that have it must by its use be valued spokespersons. The discourse of science is one of the primary controls over who can speak and who cannot speak with the authority of science.

Issues Surrounding Schooling

The question that is so rarely asked, "In whose interests is the science taught in schools?" is pertinent here. Underpinning the schooling of science is the powerful and persuasive influence of the ideology of the scientific community, and it is a unifying feature, science education everywhere. Science teachers, being certified by the institution of science's gatekeepers, gain their status and the right to speak for science. Powerful forces of convention constrain teachers to accept the ideals of the myths of science, and this creates a formidable barrier to reform. It explains why the "rhetoric of conclusions" has been the dominant feature of the way science is taught. Barriers to the reform of the schooling of science cannot be underestimated. These barriers include, for example, acknowledging and embracing its problematic nature, ensuring that students have an ability to question and weigh evidence, and developing an understanding of the social nature of science. They may only be able to be changed by the forces of national and international testing (Fensham & Harlen, 1999).

The American Association for the Advancement of Science (1989) acknowledged the importance of social responsibility as a factor that needs to be considered in developing the content of school science. Their rhetoric about this was surprisingly forthright: "likely to help citizens participate intelligently in making social and political decisions on matters involving science and technology," implying that the nature of science needs to be understood. However, their Benchmarks document (AAAS, 1993) failed to honor this imperative, perhaps because in spite of the rhetoric science is "an institution . . . dependent on the existing social order . . . and is saturated with the ideas of the dominant classes" (Bernal, 1939, p. 11).

What We Advocate

The base-line explanatory concepts required to understand the scientific arguments surrounding the social relations of science need not be taught as a rhetoric of conclusions—but historically, illustrating their essential nature as the production of explanation, not as immutable laws. They need to illustrate the strengths and weaknesses of reductionism. It is perhaps more helpful to concentrate on meta-cognitive knowledge which might be useful in tackling particular science issues within the context of the individual's own life. In other words what kind of generic skills and knowledge are required? The process of being able to participate rationally and effectively in the social relations of science is illustrated in Figure 1.

Table 1 lists the kinds of epistemic tasks in the classroom that would enhance these abilities.

The epistemic tasks which underpin these skills are by no means new, but they are usually simply listed in a vague feel-good way as things to aim for, e.g. inferring, predicting, investigating. We believe they are at the core of the work of science teaching,

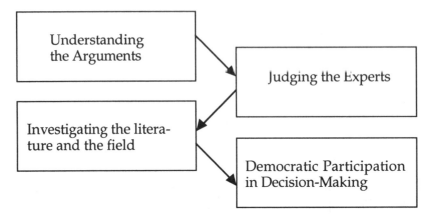

Figure 1. The abilities that lead to democratic participation in decision-making.

that they must be purposefully used to help future citizens develop strategies to answer the particular scientific issues confronting them (Cross & Price, 1992, pp. 105–107).

Structure for Project Teaching

Project teaching as we advocate it requires the following six processes.

1. *Defining the Projects*: topics are usually too broad to be successfully handled as a single project. Therefore, the first task is to work out the possible projects within it, particularly looking for those involving some scientifically as well as socially controversial issues. It is where there is controversy that the nature of science can be demonstrated and students learn the basis of making judgements. A good topic should yield three or four projects from which a teaching unit could be developed. The teacher must then decide whether to select one of these for detailed examination or whether to allot different projects to different groups of students within the class. Choice will depend in part on overall teaching

Table 1: Skills that enhance inquiry abilities

Understanding the arguments
- Understanding the processes of argumentation: to bring about acceptance of a particular proposition usually involves logic, evidence, values, beliefs and rhetoric.
- The significance of rhetoric and who makes the argument.
- Distinguishing between evidence and inference
- Distinguishing between scientific concepts and social commentary
- Seeking out the underlying scientific concepts and understanding them in context

Judging the expert
- Revealing the interests of the expert
- How the expert knowledge is used—what does the expert predict and how might that support the arguments used and why?
- What scientific concepts does the expert use?
- Devising questions to put to the experts to distinguish evidence and inference.

Understanding the arguments
- Use of libraries, the Worldwide web, and data bases
- Accessing relevant authorities and organizations
- Carrying out field investigations, measuring, mapping, recording
- Preparing reports that answer the questions, make cases, and predict scenarios

Democratic participation in decision-making
- Learning how people use the media and other forms of persuasion to press their case
- Presenting an argument—understanding the place of rational argument based on evidence and the use of rhetoric
- Knowing how to enact an inquiry
- Using democratic means of persuasion, through the Web and media
- Fielding questions to rebut counter claims

aims and on the particular knowledge and skills required by different projects.

2. *Sorting out the Questions*: this stage involves consideration of and setting out the major questions in each selected project. The scientific questions must be distinguished from and related to the

various social questions (involving ethical, political, economic and other issues) and decisions made about how to handle them. Knowledge of the particular students will assist in deciding the best approach, where the scientific, the ethical or other social questions might form a basis of interest or concern. Decisions must be made as to which are the most valuable questions for teaching purposes, which questions are too difficult for the particular class, for whatever reasons, and whether they can be handled satisfactorily in the time available. Possibilities of links with other subjects and with out-of-school agencies and activities may also be considered here.

3. *Considering the Concepts*: this stage is self-explanatory. Production of a concept map for the topic or project will enable teachers to see just how the topic relates to the syllabus, to concepts already taught or to be taught in the future. The particular concepts required for the topic could then be defined and thought given to how to teach them.

4. *Investigations*: Students undertake a variety of investigations on their project. The World Wide Web and other resources from organizations will greatly assist. Actual hands-on in the field can be undertaken whether these be in the form of scientific investigations with experiments (e.g., water quality) or social science investigations such as interviews of officials and so on. Students also evaluate documents purporting to explain the issue. They develop discriminatory skills in tabulating evidence and inference.

5. *Modeling Public Debate and Decision-Making in the Classroom*: Here simplistic debates are of little or no value. They degenerate into the use of rhetoric rather than structured argument. The process of choice is value-tree analysis (von Winterfeldt, 1986), where small teams of students prepare cogent arguments underpinning actual value positions. They do this by representing particular stakeholders. They are required to construct a value-tree illustrating how their primary value (say, fishermen objecting to inshore ocean drilling for oil—in this case the maintenance of their livelihood) affects the arguments they mount. Value trees are then place, side by side and the nature of the controversial issue across the community, and its scientific aspects are clearly seen and ap-

preciated. At no stage should students engage in litigious (and often unproductive) debates.

Exemplar: Food

Defining the Projects

As one of the fundamental biological human needs is food this has long been a favorite topic in science syllabuses. It is one in which the social and the scientific are particularly interestingly intermeshed. Eating is a form of celebration with significance ranging from the convivial to the religious. In countries like Australia where 'multiculturalism' has become a political and educational policy, food is emphasized as a distinguishing mark of cultural diversity, and enjoying different cuisines has become a recreation. At the same time our television screens bring contrasting pictures of the starving poor and the problems of the rich in disposing of food surpluses. Clearly this is an important topic for the exploration of social responsibility in science with many possible ways of approach. We suggest the following four projects: (A) Nutrition and Health, (B) The Production of Food, (C) A Sustainable Agriculture, and (D) Genetically Engineered Food production. Because this last topic is fast becoming the only topic considered in relation to food and social responsibility in science teaching we have put it aside and consider Projects A, B, and C.

Project A

Nutrition and Health: this project is closest to the most common approach in schools, that of food tests and the study of components of diets. The focus is on the consumption of food and its consequences rather than on food production, though, of course, the two cannot be completely separated. Questions of what might be called the cosmetic approach to food, the concentration on appearance rather than nutrition and natural flavor are among the topics which overlap Projects A and B. Social questions may be those of the rich countries, about healthy eating, obesity and slimming, or meat-eating versus vegetarianism; or they may be those of the poor countries and the poor in the rich ones: malnutrition of a severe kind affecting a high proportion of the population. An increasingly important question that, with industrial pollution, now concerns the rich countries is the provision of potable water. For the rich it is no longer simply the question of fluoridation. Other issues include the use of cook-chill methods of catering (Sheppard, 1987) or the use of additives as preservatives, texture modifiers, colors and flavorings and processing aids (Miller, 1985, p. 50–51). Still other issues are the irradiation of foods as a method of preservation (Webb & Lang, 1987) and the possible effects of hormones and antibiotics in the production of meat. In all these issues there are complex political economic as well as scientific questions to be teased out and evaluated.

The composition of a healthy diet is a familiar topic in the school and outside it: sugar or a substitute; salt, especially the sodium-potassium balance; fats, saturated versus unsaturated; vitamins and minerals. The socially responsible slant would stress how we know what we think we know. It might introduce such techniques as gas chromatography and mass spectrometry, opening the "black boxes" a little so as to show the problems involved in determining the composition of foods (not least, in practical terms, the cost of so doing) (e.g., Snell, 1986, pp. 139–142). It would discuss the problems of fashion, including the social determination of health and beauty (fat versus thin), and those of bias through involvement in the production and marketing of particular foods (e.g., the sugar or dairy industry). One point of entry with wide scientific possibilities is food labeling (Luba, 1985).

Project B

The Production of Food: the suggested focus of this project is on those aspects of the production and processing of food which are characteristic of modern capitalism and especially depend on the application of contemporary chemistry and biology. Production, contemporary agriculture, presents a number of social issues that lead us directly into the natural sciences. Considered from the point of view of food, pesticides (Snell, 1986) and fertilizers (Dudley, 1986) should be considered for their threat of leaving residues which might be a health hazard. Their other roles may be left to Project C below. Other residues which are causing concern include antibiotics and hormones used in the production of "modern meat" (Schell, 1984).

The processing of food involves questions of health and hygiene, many of which have already become social issues. Miller (1985) notes that additives form a potential hazard for food workers as well as consumers. Exposure at work can be through inhalation of air or by skin contact. Hazards include asthma, dermatitis, cancer, and damage to the reproductive system, depending on the substance concerned. Miller also notes that problems of microbial contamination, toxic components of natural foods and the nutritional value of foods are important questions that are inter-related with those of additives. Recently there has been interest in the water content of foods, a topic which ranges from the diluting of beers to methods of food preservation which involve water (McCrea, 1987).

Ethical questions include the treatment of animals (methods of slaughter; keeping animals in batteries during their entire life) and the export of chemicals for use in Third World countries which are banned for use, for example, in the USA. As in Project A, there are questions of bias to be elucidated and the problems of the methods used to discover and evaluate evidence.

Project C

A Sustainable Agriculture: this approach puts the emphasis on concepts of productivity, sustainability (regenerability) and questions of resource depletion and energy consumption. The stimulus to enquiry stems from two directions: the realization that petrochemicals are a dwindling resource requiring saving and replacement; and the losses of soil and water through poisoning, soil erosion and desertification, and of animal and plant species. The overall question is how to find an agriculture that will support a

high standard of living on a world scale through the use of renewable resources of energy and materials. Problems include the efficacy of definitions of efficiency in agriculture, the evaluation of
models of economic growth and development, and calculations of
the input-output ratios in terms of energy and materials. In the
following section, we explore Project C (A Sustainable Agriculture) in greater detail.

A Sustainable Agriculture

Sorting out the Questions

For those living in an urban environment choice of this project almost certainly precludes starting with an immediate social issue
such as would face those teaching in an agricultural area (soil erosion, salination) or a Third World country. The issues are large
and to some degree remote in place and time. But increasingly the
issues appear on television in one form or another, and this can
form the starting point for the project.

One such television program was "The Hunger Machine,"
originally produced for Yorkshire Television Ltd. in the UK and
later shown in Australia. In the book of the program Jon Bennet
talks about "the myths of hunger," each of which has a science
component that is relevant to our theme of a sustainable agriculture. These myths include the following (cf. Bennet, 1987, pp.
18–27).

1. *The quantity of food*: Bennet asserts that there is plenty
   of food for 6 billion people and that the real problem is
   one of access, i.e., it is socioeconomic rather than scientific.
2. *Overpopulation*: Bennet notes that some of the industrialized, rich countries have higher population densities
   than the poor countries and that the relation between
   population and malnutrition is certainly not direct.
3. *The weather*: Bennet points out that famines are not
   simply the result of natural disasters (floods,

droughts), though these may sometimes be contributing factors.

4. *The miracle of science*: Bennet argues that hunger cannot be solved through some wonder-cure like "fertilizers, pesticides and new high-yield seeds" and he goes on to question the success of "the Green Revolution."

From this it is clear that Bennet's approach puts the emphasis on the socioeconomic. But each of his "myths" can be re-examined in the light of a variety of scientific evidence. Questions can include the basis for defining overpopulation, food (nutritional) requirements for a healthy world population, problems of food storage and transport, and those opened up in myth (4). His arguments and perhaps especially that about "the Green Revolution" can be used to examine the word holism which has recently again become popular. Pimental is one agricultural scientist who uses that word, for example where he says:

> Clearly agricultural production depends on humans, land, water, energy, and natural biota and is influenced by the standard of living of society. There is a need for a holistic approach in resource management in agriculture and society. (in Dahlberg, 1986, p. 284)

One meaning of holism in relation to agriculture is the consideration of the interrelations of the three sectors identified by Bennet (1987, p. 165) as follows.

5. *Inputs*: seeds, machinery, fertilizers, etc.
6. *Agricultural Production*: the actual work that goes into farming.
7. *Post-Harvest Activities*: including storage, distribution and marketing; everything that happens between the field and the supermarket shelf.

In this context it also means taking account of the interactions of the scientific and the socioeconomic.

Among the socioeconomic questions are those about production and dissemination of information and those concerning power and decision-making in the three sectors above. The first involves questions about agricultural research, a topic that raises important

questions of particular versus general interests. The last raises the
big questions about international relations, the relations of governments and peoples to the interests of particular "multination-
als" or "conglomerates" and the nature of agrobusiness. It has
been pointed out that while on a world scale Sectors 1 and 3 are
dominated by multinational corporations, few of these like to engage in the process of farming itself. As a Tenneco spokesman put
it: "Agriculture is a high risk business and typically shows little if
any profit, especially for large corporations" (Burbach & Flynn,
1980, p. 31). As a consequence, the small peasant farmers take
the risks while the big corporations, who dictate the technology to
be employed, take major profits. At another level there are severe
conflicts of interest where traditional farming practices are replaced by commodity production, as in the so-called "Green
Revolution" areas. One example of this process, which is not part
of the "Green Revolution," is the production of severe water
shortages through the differential access to and use of water by
farmers growing cash crops, such as the chili growers in Rajasthan.
As a result of such practices many poor farmers are being deprived of water and artificial droughts have been created (Goldman, 1989).

The holistic approach also means taking account of the ethical
issues involved. This is the subject of an article by Freudenberger,
"Value and Ethical Dimensions of Alternative Agricultural Approaches," where he sets out four questions about "the values and
goals of both agriculture and society." (The following are quoted
in Dahlberg, 1986, p. 349.)

1. Can social and agricultural systems be developed that
   stress the importance of responsible freedom in building a responsible society?
2. Can social and agricultural systems be developed that
   provide meaning in one's work and in one's relationships with society and natural environment?
3. Can social and agricultural systems be developed that
   recognize the primacy of life and its pattern of suste-
   nance (in a full ecological sense)?

4. Can social and agricultural systems be developed that recognize the need to include the welfare (health, stability, integrity, beauty, harmony) of future generations of both human as well as non-human life forms?

These questions could make a useful start for discussion, but the undefined abstractions like "responsible freedom" would require defining, and they must be related to specifically scientific questions. Freudenberger himself offers a definition when he further asks: "How can a society be judged responsible if it cannot be sustained independently of external and/or non-renewable resources?" (Dahlberg, 1986, p. 352). This question takes us immediately into the scientific with mention of (non-)renewable resources. He also raises relations between the industrialized countries and the Third World when he says: "Can a society consider itself to be just if it is sustained at the expense of other societies in other places?" Other abstractions on which science teachers can throw light are "normal" and "natural." Freudenberger uses "normal" in speaking about "radiation patterns of incoming light through the ozone shield" and "flows of oxygen, carbon dioxide, and nitrogen generation and absorption" without stating the time period he has in mind. Similarly, he uses "natural" to refer to the survival of such threatened species as elephants, whales and buffalo (in Dahlberg, 1986, pp. 358–359).

Finally, in sorting out the questions it is important to be clear about the central question, the concept of sustainability. Freudenberger prefers the term regenerative agriculture, which he defines as follows.

> A regenerative agriculture is generally conceived as one in which the number of people, the rate of use and regeneration of essential resources, and the rate of waste production are within the capacity of the earth to support and absorb, and where an acceptable quality of human life can be sustained indefinitely. (p. 358).

Regenerative agriculture is preferred to "sustainable agriculture" because the latter allows for high output and profit to be continued for a long period through the use of such inputs as cheap but non-renewable fossil fuels. Conflicting points of view are described by Tisdell who notes that "some economists as opposed to some ecologists see no particular virtue in a sustainable economic system" and that the issue is "complex" (Tisdell & Maitra, 1988, p. 36).

Another approach to sorting out the questions could be to examine the "developmentalist myths" which Levins (1986) describes and then shows why they are myths. They are:

1. Backward is labor-intensive, modern is capital-intensive;
2. Diversity is backward, uniform monoculture is modern;
3. Small scale is backward, large scale is modern;
4. Backward is subjection to nature, modern implies increasingly complete control over everything that happens in the field, or orchard or pasture;
5. Folk knowledge is backward, scientific knowledge is modern;
6. Specialists are modern, generalists backward; and
7. The smaller the objects of study, the more modern they are.

Handling the Arguments

It would seem that there are four arguments that are particularly pertinent to the questions of a regenerative agriculture: the input-output balance of energy; the maintenance of genetic variety; organic farming methods; and the so-called Green Revolution. The last provides a useful lesson in the history of science.

1. *The input-output balance of energy*: One of the central issues is that of the energy input-output ratio and the dependence of "modern" agriculture on using large quantities of non-renewable oil. The history of this discussion forms the core of Ecological Economics (Martinez-Alier & Schlüpmann, 1987). Among the important considerations they raise is (a) that "modern" agriculture uses up much more non-renewable energy than it supplies in the product; (b) the highly industrialized countries use much more non-renewable energy for non-agricultural purposes than they do in agriculture. They are therefore not arguing that poor peasant

agriculture should be forcibly maintained by the rich countries in order to serve the latter's energy needs (pp. 241–242). This is a topic that opens up a number of scientific questions. These questions focus on diverse topics such as human energy requirements, energy provision by different kinds of food, solar energy consumption by plants, energy inputs of different kinds of agriculture (hand labor, animals, tractors), energy used in the feeding of farm animals and the production of farm tools, and the possibility of different energy sources, renewable and otherwise.

2. *The maintenance of genetic variety*: The commercializing of biology which is implied in the terms agricultural biotechnology and genetic engineering depends for its success on the ability of the private firms being able to control "their" products through patenting and other devices. Hybrid seeds gave them this control through biological means (Berlan & Lewontin, 1986). The patenting of plants has a history going back into the last century (Mooney, 1983, pp. 134–166). Already in 1930 the American Congress adopted a Plant Patent Act, but this only included asexually propagated plants. The Nazis established Plant Breeders Rights in Germany, Austria and the Netherlands in the thirties and forties, and the policy has been taken up in the European Common Market and again by the USA in recent decades. As well as the moral, legal and economic arguments, there are important scientific questions. These in the main concern the importance of maintaining genetic diversity. Suzuki and Knudtson (1988) have formulated this as one of their "Genethic Principles": "Genetic diversity, in both human and nonhuman species, is a precious planetary resource, and it is in our best interests to monitor and preserve that diversity" (p. 290). He illustrates this principle with a discussion of maize. Mooney (1983) takes a wider canvas and discusses the role that traditional varieties ("land races") and wild relatives of cultivated plants play in the plant breeding process.

Maintenance of genetic diversity is one of the major scientific aspects of a sustainable agriculture. It involves such qualities as yield (size of grain, etc.); straw yield (dwarf versus other varieties); pest resistance; drought resistance; and nitrogen fixation. Modern genetic techniques raise important questions about the

Table 2: Concepts and subconcepts

Structure of matter:
 atomic structure, protons, neutrons, and electrons
 elements and isotopes
 compounds and mixtures

Diffusion and solubility

Organic and inorganic compounds

Energy:
 transformation and conservation
 mass-energy transformations
 sources, renewable and non-renewable
 energy flow and balance

Machines:
 types and efficiency

Biology:
 tissues, cell structure, chromosomes and cell division, genes
 cancer
 use of isotopes in biological research
 photosynthesis
 hybrids, gene pool, pure lines
 control of pests (biological and chemical)
 plant growth, nutrients, nitrogen and phosphorus cycle

Geology/geography
 atmosphere
 hydrosphere, water cycle, clouds, rain, aquifers
 weather
 rock types, natural radioactivity
 temperate and tropical climates

possibility of artificially improving genetic diversity (Abelson, 1975).

3. *Organic farming methods*: One of the controversies connected with soils and cropping practices is that of organic versus artificial fertilizers. Here the issues are not as clear as might at first sight appear. The central issues are clearly fertilization and pest

control. But, as Lockeretz points out, more general principles are involved: sustainability, working with nature, healthy soil, ecosystem compatibility and a holistic approach (in Dahlberg, 1986, p. 295). Use of one thing, for example, large tractors for weed control rather than toxic herbicides may conflict with other principles, such as reducing dependence on non-renewable petroleum products or off-farm inputs. Many farmers are practicing reduced-chemical or partial organic farming (Brusko, 1985). Contributing to regenerative agriculture is, of course, different from arguments about whether the products of organic farming are healthier (contain less toxic residues, etc.). Those questions belong to another project. The connection of organic farming and control of soil erosion is an important issue. It is interesting to note the support given to organic farming through the Agricultural Productivity Act of 1983 by the US Congress in the mid-eighties. Lockeretz notes the combined effect of soil erosion, grain surpluses and pesticide contamination on possible future federal government policy (in Dahlberg, 1986, p. 299). Revelle refers to problems and possibilities in Africa (Simon & Kahn, 1984, p. 191). Discussion of all these issues is a nice exercise in sorting out arguments and measurement of variables. It also provides a useful example of the complexities of practice compared with the laboratory-style scientific test.

Considering the Concepts

Science concepts involved are listed in Table 2 (p. 116).

Particular Teaching Methods

The Bangalore Conference discussion on the topic of food, agriculture and education recommended a number of teaching approaches. These included "investigatory approaches such as the building up of statistics on national consumption of certain foods; sugar and incidence of for example, diseases such as heart diseases/diabetes" and "field trips to food processing industries and centres using biotechnology for food production" (Rao, 1987, p. 189). The workshop on *Land, Water and Mineral Resources in Science Education* of the same conference has some nice examples on water resources, including how to organize a role-playing game on Saving the Rhine from Pollution (Graves, 1987, p. 183). Throughout these volumes the emphasis is on linking theory with practice based on local conditions (Rao, 1987, p. 189–190, 197, 211).

Conclusion

We believe that the prevailing model of science, now widely criticized as positivist or reductionist, needs changing if science is to fulfil scientists' long held belief in its ability to contribute towards human betterment and happiness. By concentrating on single causes and neglecting complex interrelations, especially in biology and other higher-level sciences, the real nature of phenomena is not grasped. We also believe that such change is a necessary part of orienting science teaching towards an emphasis on social responsibility and thus towards a science, particularly in its technological aspects, which can guarantee a sustainable future for humankind.

As part of the attempt to teach science in an anti-positivist, anti-reductionist way it is both necessary and possible to teach it in a more theory-conscious way. That is, to make clear the role of particular types of theory in the scientific process, whether different forms of explanation, the influence of an explanation on particular observations and their interpretation, or the influence of explanation on theory as know-how. We are strongly opposed to what sometimes appears to be the case, the simple replacing of one "rhetoric of conclusions" by another.

While we see it as necessary for social responsibility that the teaching of science as a whole changes in this way, we do not see that as sufficient. Nor is it likely to happen sufficiently quickly to meet immediate needs. Particular social issues require understanding of particular pieces of science. Particular explanations may be in question, and particular skills may be involved in the evaluation of the evidence concerned. Emphases on the knowledge and skills required for making judgements and particularly for the handling of evidence will help ensure future citizens will be able to handle the issues that affect their lives as and when they occur.

The approach to science teaching that we are advocating here is perfectly consistent with the liberal democratic tradition in education. We are concerned with the public good and see as part of the critical scientific attitude the elucidating of the values, attitudes and interests that are embodied in science as Institution. Indeed, this answers the question: "In whose interests do we teach science?" We make no pretence of being value-free, nor, as we have stressed, is that possible. That still-expressed claim is increasingly widely recognized as an illusion. What we have stressed is the importance of detecting and judging the role played by the values that are involved in science. We would argue that bias is not necessarily a bad thing. Bias in favor of preserving biological species, or avoiding war, may be good. The important thing is to recognize what bias exists. At the same time one should recognize that bias is not itself evidence and says nothing about the truth or falsity of the point at issue.

While our approach is open on the question of teaching methods, welcoming exploration of variety depending on circumstances, we have expressed some reservations on discussions and debates that may bear repetition here. We are skeptical of the value of the "for" and "against" method of debate which occurs in some classrooms, where there is more heat than light and opinions are asserted rather than examined in the light of their grounds and backing.

To repeat, it is not viewpoints but their grounds and backing which need examination. At the same time, the use of the word "balance" all too often suggests simply two positions, e.g., the "two extremes" mentioned here. Together with the word "extreme," this introduces a prejudice that we believe our concentration on the formulation of a variety of questions does not. Considerations like these suggest that small group discussion of question formulation and the evaluation of evidence may in most cases be a more fruitful teaching method than whole class discussions or debates—which is not to rule these out when circumstances are right.

Throughout this chapter we assumed that the threats to our world are various and serious, but not that in particular cases the evidence is conclusive. A future has been put in question and given certain conditions, such as a nuclear war or the worst case scenario of the ozone layer, the outcome would appear certain. Our approach is to examine the evidence for different scenarios with a view to recommending the conditions under which a sustainable future is possible. It remains to recommend what to do in those cases where a scientific judgement is not possible, where the balance of evidence is insufficient to make a recommendation on scientific grounds. Here one must step back from pretensions to science and turn to caution and common sense. If there is reasonable evidence to suggest that action will lead to unacceptable outcomes then the decision must be to refrain. The problem remains: what is "reasonable"?!

In recommending a socially responsibile approach one faces the objection that problems are too big and depressing and that it will turn students against science rather than motivate them to study it further. This is a justified fear and social issues must be carefully chosen and treated to take it into account. If possible at least a few of the topics must be ones where it can be shown that a positive outcome is likely and that student consideration and, where possible, activism of an informed and considered kind, can assist in this. Topics like the handling of waste locally, the preservation/conservation of some species of animal or plant, or the monitoring of some hazard or other which is causing a problem locally are good examples.

Finally, while science teachers, other than those in the tertiary education sector, are seldom in a position to carry out fundamental monitoring or research into the science of social issues, they are in a potentially strong position to exert an effect. Numerically they are a large group, in terms of education they are a well-informed group and they are also a professionally organized group. They not only are relatively free from the vested interests which today hamper many research scientists from speaking out, but they have a professional interest in the welfare of the future generations, and therefore in the science for a sustainable future. Conscious of this power and convinced of a way to exercise it, their influence could be far-reaching.

References

Abelson, P. H. (1975). *Food: Politics, economics, nutrition & research.* Washington, D.C.: American Association for the Advancement of Science.

American Association for the Advancement of Science. (1989). *Science for all Americans: Project 2061.* Washington, D.C.: AAAS.

American Association for the Advancement of Science. (1993). *Benchmarks for science literacy.* Washington, D.C.: AAAS.

Bazelon, D. L. (1979). Risk and responsibility. *Science, 205,* 277–280.

Bennet, J. (1987). *The hunger machine: The politics of food.* Cambridge, England: Polity Press.

Berlan, J.-P., & Lewontin, R. C. (1986), The political economy of hybrid corn. *Monthly Review, 38,* 35–47.

Bernal, J. D. (1939). *The social function of science.* London: George Routledge.

Brusko, M. (Ed.). (1985). *Profitable farming now.* Emmaus, Pa: Regenerative Agriculture Association.

Burbach, R., & Flynn, P. (1980). *Agribusiness in the Americas.* New York: Monthly Review Press & North American Congress on Latin America.

Cross, R. T., & Price, R. F. (1992). *Teaching science for social responsibility.* Sydney: St. Louis Press.

Dahlberg, K. A. (1986). *New directions for agriculture & agricultural research: Neglected dimensions & emerging alternatives.* Totowa, N.J: Rowman & Allanheld.

Dudley, N. (1986). *Nitrates in food & water.* London: London Food Commission.

Fensham, P. J., & Harlen, W. (1999). School science and public understanding of science. *International Journal of Science Education, 21,* 755–763.

Goldman, M. (1989). The "Mirch-Masala" of chili peppers: The production of drought in the Jodhpur region of Rajasthan. *Capitalism, Nature, Socialism: a Journal of Socialist Ecology, 2,* 83–92.

Graves, N. J. (1987). *Land, water & mineral resources in science education.* Oxford: Pergamon Press.

Levins, R. (1986). Science and progress: Seven developmental myths in agriculture. *Monthly Review, 38*(3), 13–20.

Luba, A. (1985). *The food labelling debate.* London: London Food Commission.

Martin, B. (Ed.). (1996). *Confronting the experts.* Albany: State University of New York.

Martinez-Alier, J., & Schlüpmann, K. (1987). *Ecological economics: Energy. environment & society.* London: Basil Blackwell.

McCrea, D. (1987). *Opening the floodgates: A report on the excessive use of water in food.* London: London Food Commission.

Miller, M. (1985). *Danger!—Additives at work: A report on food additives; their use & control.* London: London Food Commission.

Mooney, P. R. (1983). *The law of the seed: Another development & plant genetic resources.* Development Dialogue: 1–2 (Special issue).

Rao, A. N. (Ed). (1987). *Food, agriculture & education.* Oxford: Pergamon Press.

Rotblat, J. (1981). Hazards of low-level radiation—less agreement, more confusion. *Bulletin of the Atomic Scientists, 37*(6), 31–36.

Schell, O. (1984). *Modern meat: Antibiotics, hormones, and the pharmaceutical farm.* New York: Vintage Books.

Sheppard, J. (1987). *The big chill: A report on the implications of cook-chill catering for the public services.* London: London Food Commission.

Simon, J. L., & Kahn, H. (1984). *The resourceful Earth: A response to Global 2000.* Oxford: Basil Blackwell.

Snell, P. (1986). *Pesticide residues & food: The case for real control.* London: London Food Commission.

Suzuki, D., & Knudtson, P. (1988). *Genethics: The ethics of engineering life.* Sydney: Allen & Unwin.

Tisdell, C., & Maitra, P. (1988). *Technological change, development & the environment.* London: Routledge.

Webb, T., & Lang T. (1987). *Food irradiation: The facts.* Wellingbor-
    ough, Northamptonshire: Thorsons Publishing Group.
von Winterfeldt, D. (1986). *Value tree analysis: An introduction &
    application to offshore drilling.* In P. R. Kleindofer & H. C. Kun-
    reuther (Eds.), *Insuring and managing hazardous risks: From
    Seveso to Bhopal and beyond* (pp. 349–377). Berlin: Springer-
    Verlag.
Ziman, J. (1980). *Teaching and learning about science and society.*
    Cambridge, England: Cambridge University Press.

6

Beyond the Bold Rhetoric of Reform: (Re)Learning to Teach Science Appropriately

Kenneth Tobin

It is so easy to speak strongly about what ought to be done in science classrooms and so difficult to match words with appropriate actions. This is the crux of the reform problem in science education. For as long as we have had formal education it seems as if we have had well-intentioned others wanting to enact reform (e.g., Rutherford & Ahlgren, 1990). Recommendations have taken many forms but usually have several common aspects: those with power prescribe changes for others with less power, and the changes recommended usually apply to all teachers and students within a system (e.g., "all standards for all learners" [National Research Council, 1996]). I have done my fair share of exhorting others to enact science curricula in improved ways (e.g., Tobin, 1987). However, recently I have experienced first hand the enormous difficulty of teaching science in urban schools in ways that make a difference to the lives of students (Tobin, 2000). In this chapter I want to draw attention to a folly that has permeated exhortations of reform in myriad national-level reports that followed the wake-up call embodied in *A Nation at Risk* (National Commission on Excellence in Education, 1983). This folly is to present goals for science education as master narratives that do not acknowledge the centrality to teaching and learning of science of issues concerning who, where, what, when, and how. It can be exhilarating to draw attention to the plight of the underdog and exhort science for all. However, it is folly to take such exhortations too seriously unless there is a conscious awareness of what it takes for those in the front line who are expected to catalyze successful reform initiatives (Rodriguez, 1997). Furthermore, serious ontological issues must be addressed in translating recommendations for reform into enacted curricula and higher achievement for all (Roth, 1998; Roth & Tobin, 2001a, 2001b).

Ontologically there is a gap between the exhortations to reform science education and enacting those reforms. Recommendations to change are themes selected to re-present what would-be-reformers have identified as desirable states for communities in which reform is specified as necessary. As Roth (1998) has shown, there is an ontological gap between actions and re-presentation, a gap that teachers encounter every day as they endeavor to enact activities as planned. It is one thing to exhort differences in the ways teachers plan for their teaching and quite another for them to teach as they have planned. Roth points out that effective teaching involves appropriate action in contexts where usually there is insufficient time to think about what is and is not appropriate (particularly if teaching is regarded as cascades of contingent actions).

Voiceover: I do not mean to imply that only politicians and policy makers exhort others to change their practices. On the contrary, most researchers over-generalize their findings (including myself in numerous papers). Teacher educators are notorious for prescribing master narratives that are grounded in research and implicit theories that are highly valued (frequently without taking the time to show how these master narratives can be applied to obtain the putative outcomes).

Teaching as Praxis

Teaching is a form of praxis in which unfolding events in a community *set* a context in which the *habitus* of teachers (Bourdieu, 1992; Roth, 1998; Tobin, 1999) generates actions, without reflection, to address contingencies that unfold with time in response to the specifics of classroom events and phenomena. A teacher's *habitus* is a set of structured dispositions built on the experience of having taught. *Habitus* does not reveal itself in what a teacher might say or write about teaching, but becomes evident in particular situations when teaching occurs (i.e., in praxis). Accordingly, while teaching, a person's practices are shaped by the actions of the participants in a community and the *habitus* of the teacher. The concept of *Spielraum* also is important in making sense of teaching (Roth, Masciotra & Boyd, 1999; Roth, Lawless, & Masciotra, 2001). *Spielraum* allows a teacher to act appropriately at given times and, through action, to anticipate problematic situations. As a teacher gains experience in a given context s/he builds *Spielraum* that provides room to maneuver and possibilities to act without consciously attending to her actions (Roth & Tobin, in press-a). What is not captured, and cannot be encapsulated in exhortations for reform, are the pathways to afford the rhetoric by generating essential *habitus* and *Spielraum* for teachers.

I do not argue here that recommendations for reform have no value but that they are difficult to enact, particularly when presented as one-size-fits-all panacea for specific communities. It is not sufficient in recommendations for reform only to describe the differing contexts that pertain to stated recommendations. More significant is to acknowledge the centrality of context in teaching as praxis, a form of practical action, in which the next move always is located in unfolding events that cannot be predicted a priori.

My Goals for Teaching Science in an Urban High School

I knew what to do when I decided to teach science in a large urban school in Philadelphia. After all that is why I came to the University of Pennsylvania to continue my career as a science educator. I wanted to work with groups of students who were underrepresented in science by virtue of their social and cultural histories. City High School, in the western part of Philadelphia, has a profile that is appropriate for pursuing my goal of making a difference through science education to the lives of students from urban communities, especially those from conditions of poverty. Increasingly schools like City High are associated with low academic achievement and feelings of helplessness about the trans-formative potential of education (Mirón & Lauria, 1998; Toussaint, 1997).

My initial visits to science classrooms at City High confirmed my worst fears. The curriculum was enacted at a minimal attainment level, students rarely engaged appropriately in activities; equipment, supplies and textbooks were in short supply, and there appeared to be a lack of motivation on the part of either teachers or students to pursue deep learning goals (Tobin, Seiler & Walls, 1999). Conversations with teachers revealed that they placed the blame for this state of affairs on the students and the communities in which they lived. Teachers also noted a lack of commitment from the school district and a system that permitted urban schools to be funded at a level far below that of suburban schools. In a striking contrast the students placed the blame for the inadequate curriculum squarely on teachers and administrators who maintained a curriculum perceived by many students to be a complete waste of their time.

I was determined to enact a science curriculum that was liberatory and potentially transformative. It is characteristic for such curriculum that, by virtue of their participation and learning, students would have a better appreciation of their world, enhanced opportunities for advanced study in science, and increased choices for career placement and training (Barton, 1997). I wanted to enact a curriculum that the students would perceive as interesting, relevant to their lives, and useful. To the extent possible I wanted the students to have choices in what they would study and where they would study it. I predicted that they would enjoy doing science if the program was based on investigations, and I had a preference for the activities to involve real world problem solving. I wanted to focus on inquiry as a means to develop deep understandings of science subject matter. The critical defining characteristic of my approach to the science curriculum would be deep learning whereby students would pursue areas of interest in detail and, in so doing, employ a multitude of resources to support their learning.

Characteristics of the School

City High School has about 2,300 students arranged in 10 small learning communities (SLCs) that can be considered as schools within schools. Each SLC has its own students, classrooms, and teachers. The idea is that students stay together with teachers over the duration of their high school lives and greater personalization of the curriculum leads to higher achievement and a feeling of *esprit de corps* as a result of belonging to a SLC and getting to know the students and teachers. Two of the 10 SLCs have a curriculum intended to be college tracked whereas most of the others are somewhat career oriented. For example, students in *Health* often pursue health-oriented themes in their studies, and they regularly undertake field studies in health institutions. However, one SLC caught my attention because it was different than the others. *Opportunity* is for students with histories of failure at, and who have not settled into, high school. This SLC appeared to be the most disruptive. Numerous factors, such as sporadic student attendance, seemed to make it virtually impossible to enact a coherent curriculum. From my perspective the school abandoned the students in *Opportunity*, and I felt a strong commitment to identify ways to provide them the opportunity that was so cynically promised by the label for their SLC.

When I announced my plans to teach in *Opportunity* I could sense from the reactions of teachers, administrators, and colleagues that few considered it a worthwhile investment of my time. There was an explicit message that I should allocate resources in ways that would be of benefit to students in the other nine SLCs. Such reactions were consistent with my impression that *Opportunity* was created as a dumping ground for misfits. It was a place to hold 200 students for the benefit of the other 2,000 students in the school who could then work without having to contend with the constant disruption of an unsettled and often alienated minority. Consistent with the notion of *Opportunity* being a holding site for misfits was my perception that fewer resources were allocated to this SLC and there was less of an institutional concern for a high-quality curriculum.

Planned Curriculum

Even though I began to teach with a transformative/liberatory agenda I was by no means sure of how to proceed with the teaching of consecutive units on chemistry and physics. I intended to use a multi-faceted approach. For example, I wanted students to read about contemporary science and thought they could access science from journals, newspapers, magazines, and the Internet. To accommodate my initial thoughts on what the enacted curriculum might be like I requested that my class be scheduled for two days in a computer lab and three in a science room that would support a range of activity types, including investigations.

I planned to begin with an activity sequence on chromatography, examining the dyes from M&Ms and Skittles because those items of candy were potentially interesting to students, involved familiar materials, and could easily connect to a unit on food and nutrition. While studying food coloring and the relative safety of different dyes I believed we also could study the chemical constituents of the foods consumed by students in various meals. As part of a unit on food and nutrition I considered that students might grow sprouts, radishes, and other edible plants. Also, within the chemistry course, I thought they might grow fast plants, study life cycles, and learn how to grow nutritious plants, such as corn, in relatively quick time. I did not expect that these activities would replace chemistry but set a context in which concepts of chemistry could be explicated.

Barton's research with the homeless from New York (Barton, 1998) encouraged me to think that I could enact a curriculum about which my students would have passion. Accordingly, I planned an activity sequence on *Street Science* that was ready to go as soon as I perceived an appropriate opportunity to enact it. I had been exploring my neighborhood and, with a digital camera, had captured images to focus discussions on science-related topics. My purpose in doing this was to show students that my own city neighborhood had aspects that were a source of personal dissatisfaction. A discussion of the slides could lead to the identification of other aspects of city life that lend themselves to scientific study. I was aware of the dangers of focusing only on the negative aspects of an urban neighborhood but wanted to generate a list of potential questions for study. Even though I considered questions such as, "What is it that stinks?" to carry a negative connotation I regarded it as a potential springboard for inquiry. For example, students could explore a variety of smells, find possible underlying causes, and identify ways to prevent or remove them. Similarly, garbage dispersion and disposal were critical issues that detracted significantly from the ambiance of my neighborhood. I regarded such topics as having considerable potential to connect to important social issues that might result in changed policies being considered by agencies involved in garbage collection and disposal. I could imagine how issues like the packaging of foods and the marketing of fast foods also could connect a unit on garbage to one on health and nutrition. Incineration of wastes and recycling also were topics that could be developed if student interests supported such activities.

Making Science Relevant

Each morning I gave students an activity sheet containing questions to review from the topics we had dealt with the previous day. Of the 35 students in my class fewer than five could be relied on to provide answers that came close to being correct. One morning, because of my frustration at the students' blank faces and their unwillingness to interact with each other or me, I decided to change from activities concerning the periodic table of elements to my planned unit on *Street Science*. The protests of students were immediate. I was caught completely off guard when previously apathetic students refused to accept the shift in focus. They did not want to talk about garbage in the streets and possible ways to improve disposal. Nor did they want to discuss smells, good or bad, from their neighborhoods or the city streets. "What's this got to do with the periodic table?" I was startled that any student would even try to connect the discussion I had initiated with what we had previously been discussing, and I did not have the management skills to respond to questions that challenged my authority. The quick turn of events exposed the lack of rapport I had established with these students and emphasized that I had not yet earned their respect.

Voiceover: If only I could see then what I see now. The involvement of the students here was liberatory. They were engaged and used their language resources to initiate arguments they found compelling. The pace of arguments was fast and interactive. The students understood others' arguments, built logically on points made by previous speakers, and used overlapping speech to maintain control. Because I could not appropriate their language I could barely get a word in without shouting or using some other move to disrupt the unintended activity that was in progress. Ironically, this may have been an opening that was for me akin to finding the Holy Grail. The students were participating and using their discursive resources or their funds of cultural knowledge (Lee, 1999) to make persuasive points.

Suddenly the students were alive as they rejected my efforts to discuss science in its social contexts. They were gaining control and I did not want that process to go too far. I eyed the clock. I had 35 minutes still to go, and the students were unruly and likely to become more so. I edged my way to the rear of the classroom where Spiegel (the regular science teacher) was seated. "What do you think we should do next?" I asked with my mind furiously reviewing the options. "Stick with it for now," he suggested. "No way!" I thought as I turned back to the class. I needed to get the students engaged again, and I was concerned that some of them might leave the classroom before it was time to do so. At that moment I felt vulnerable and needed to act decisively. Defiantly I turned to the chalkboard and resumed the initial activity on electronic structure and ionic compounds. As I wrote on the board order returned as the students assumed their accepted and familiar roles. Most began to copy; some switched off again, and I breathed a huge sigh of relief. They were compliant. On this occasion I managed to cap an energetic outburst that was beyond my ability to control.

Voiceover: These students are extremely good at debate in that they are logical and articulate. I wonder where the activity would have ended up if I had followed Spiegel's advice to proceed for now. My concern for maintaining control and order was paramount at the time, and although it is romantic to consider that I might have proceeded, it is most important to maintain safety, and the type of environment that was evolving was too heated to allow the whole class argument to continue. The essential issue here is not so much that an activity on Street Science *was not successfully enacted but that I did not have the cultural knowledge needed to successfully enact a curriculum that met the needs (as I perceived them) of the students.*

Later in the semester I again endeavored to initiate a set of activities on science, technology, and society by focusing on the students' neighborhoods. This time I prepared a worksheet entitled *Life in my 'hood* consisting of 41 questions, each of which was intended as a potential springboard for a longitudinal study that connected the study of science to aspects of their neighborhood. My purpose in using the worksheet was to provide a focus for the students to comment on issues that could later be used for in depth study of science contextualized in their lives at home.

Voiceover: Many of the students resented my use of the term 'hood. They use the term in their exchanges when they refer to their neighborhoods. My intention was to use of the term in a process of building a shared language. The students' heated objections deterred me from using the term publicly in the future. Their objections also are an example of them using issues like my inappropriate use of terms to impress their peers with public criticism of me. Later I address this issue in detail.

I thought the worksheet would provide a context for discussions about science in the home and neighborhood. I hoped to identify areas in which the students were interested. However, the activity supported my earlier suspicions that the students had difficulty in responding to open-ended questions. The small percentage of students that participated in the activity answered most questions in a word or two and did not use them as opportunities to show what they knew or wanted to know. The students did not use the questions as springboards for inquiry, and once they had responded to them they regarded the activity as complete and expected to be told the right answers. Instead of the worksheet being a springboard for several weeks of work, most students accomplished all they intended within 15 minutes. Shaneka's responses to one of the questions provide insights into the issues associated with my attempts to link science to the home and neighborhood:

Question: What parts of the 'hood could be improved?
Answer: All of it. We have a police station and fire station and so
           much stuff happens. People get robbed all the time. Houses
           stay burning down.

I made two more attempts to catalyze interest in *Street Science*. In the final three weeks of the term I began to bring one-minute readings to class. I looked through a number of books to identify readings about science that also related to the environment. These included articles on waste disposal, nuclear wastes, uses of drugs to enhance athletic performance, the presence of lead in the paint on endorsed athletic shoes, and rubber from tires as a source of air and street pollution. My strategy was to hand out the readings at the door and ask the students to read them and respond to the questions I had prepared. As the students sat down I moved from individual to individual, encouraging each to get started and to complete the work. Once again most students did not respond positively to these activities and declined to write responses to the questions.

Voiceover: On any day 4–6 students was the maximum I could get to participate. I did not believe it ought to have been so difficult to get students motivated to learn. They did not participate in debates over issues such as those I regarded as controversial. The debates they participated in, and seemed to enjoy most, concerned the relevance of the curriculum. For the most part the students did not regard activities associated with Street Science *as legitimate science, and they protested about the relevance of the curriculum whenever I gave them the chance.*

Since I had made numerous attempts to emphasize connections between the home and science I decided to put several questions on the test. (For example, "Why is it not a good idea to burn trash or bury it in the landfill?" and "What are some alternative ways of dealing with litter in the city?" [5 points].) However, the responses to the test questions were minimal and reflected my previous efforts to engage students in meaningful discussions that involved them giving input and doing research from books and the Internet. I concluded that the students need much more structure than I gave them, and their discursive resources did not allow them to employ a question as an opportunity to show what they knew about a topic or to focus inquiry and build knowledge.

Voiceover: Although I would have denied it strenuously at the time I was basing curricular decisions on deficit ways of thinking about what students knew and could do. In my search for what they were good at I was blind to the awesome fluency of their oral language and the extent to which they could bring logic to their arguments. Students demonstrated a science-like discourse in their use of evidence to argue against doing the activities I prescribed for them. And my labeling of their actions as resistance led me to think about events in ways that distracted me from a search for ways to employ their funds of cultural knowledge in class.

Learning the Culture of My Students

Why did my efforts to teach *Street Science* end in frustration? My voiceover comments allude to my inadvertent use of a deficit model for the students and general ignorance of the social and cultural attributes of the students I was endeavoring to teach. In an effort to deal with my own lack of cultural knowledge I began an ethnography of the streets of Philadelphia (Tobin, 2000). I read about other studies of African Americans in Philadelphia (Anderson 1990, 1999) and recruited Tyrone, an African American student from *Opportunity*, as a co-researcher and a teacher educator. In this section I provide some background on Tyrone and connect salient issues to my teaching of science in *Opportunity*. Tyrone's involvement in my ongoing research provided valuable insights into the culture of African American youth from West Philadelphia and also addressed issues of agency (e.g., Mirón & Lauria, 1998). He was able to express his perceptions of City High and its teachers and students being constrained by my agenda. The perspectives expressed in this paper and others that have involved Tyrone (Tobin, 2000; Tobin, Seiler & Walls, 1999) re-present issues that are of importance to Tyrone, and he has read the papers and agrees with the perspectives attributed to him. The insights provided below are based on one year of regular interaction with Tyrone in settings that included school, my office, restaurants, and the streets.

It is a year since I met Tyrone for the first time. I well recall the first day. He had volunteered to participate in a service learning course in which he would teach science to a small group of middle school students. I was coteaching (Roth & Boyd, 1999; Roth, Masciotra & Boyd, 1999) the course with Spiegel (Tobin, 1999, 2000; Tobin, Seiler & Smith, 1999). As Tyrone walked into the room I found myself instinctively backing off. Tyrone is an African American male of about 16 years of age who presents himself as a street kid. To be honest he looks like trouble. His dreadlocks are arranged asymmetrically; his demeanor is gruff, and frequently his dress is unkempt. Spiegel, who knew him well, explained to me that Tyrone is a good kid who looks like a thug and interacts with others according to the "code of the street" (Anderson, 1999). Accordingly, Tyrone is often in trouble with other teachers and non-teaching assistants whose job it is to maintain order in the school. Also, his explosive temper often lands him in trouble and leads to his suspension from school. Tyrone is cognizant of his appearance causing problems in the school. He commented that:

> Lot of people tell me it's my hair. But I'm not cutting it for no one. I don't understand it. I should be able to wear my hair, or dress whatever how I want. As long as I got my mind, right? It should be cool. It's like people just look at me automatically they just start stereotyping me. You gotta stay on him. He ain't going to do what he got to do. Stay with him. That sometimes really make me mad. It's like I'll be trying my hardest, and people just still down on me.

Tyrone comes to school when it is convenient. He walks to City High and prefers not to do so in inclement weather. He does not like detention, and so, if he might be late and thereby risks detention he does not leave home, or he does not attempt to enter the school building. Also, since his mother sleeps during the day and his father neither speaks to him nor cares whether or not he attends school, Tyrone just goes back to sleep if he awakes feeling tired or in a bad mood. Some days he just decides to hang out with friends from his neighborhood. On such occasions he may roam the streets in the northern part of the city, closer to his home than the school. Also, for a variety of reasons, Tyrone gets suspended for five days at a time and, because his mother must come with him to school to get re-admitted, he often remains away for several days beyond the 3–5 days of suspension. Not surprisingly, his relatively low rate of attendance at school has consequences for his participation, performance, and learning.

When Tyrone is at school he prefers to work and sits himself off to the side so that he will not get distracted by others or implicated in trouble by being near to students who may be off-task or disruptive. However, because of his poor record of attendance he needs to be updated by his teachers, and he expects them to provide him the necessary assistance. If teachers do not make every effort to teach him and assign work that he has missed, Tyrone becomes disgruntled, and his short temper often catalyzes confrontations with them.

Tyrone has been more than three years in grade 9. As I worked my way through his academic transcript, I was astonished at its complexity. He has enrolled for more courses than it would take to graduate (27.5 credits), but to this point in time Tyrone still has only 9.5 of the 21.5 credits needed for graduation. For example, since 1997 he has enrolled for 5.5 credits of science but has only passed 1.5 credits, all in 1999. Tyrone's infrequent attendance leads directly to his failure in many courses, and it also makes it difficult for him to understand subjects like mathematics and to gain a passing grade in any of his subjects. The consequences of this are that he needs to repeat many of his classes, leading to boredom and a sense that the teachers are wasting his time. "Have you seen this transcript before?" "No," Tyrone replied. "I have been trying to get a hold of it for some time but have not been able to." I grimaced as we worked our way through what he still had to accomplish in order to graduate from high school. Then the futility of ever completing his program of study hit me. Why would Tyrone bother to come to school?

Voiceover: Tyrone has much to say to teacher educators in terms of how to teach students like him. I look to his wisdom, and as I conclude this chapter, I find myself searching for remedies to the situation. I find myself in awe with respect to his own learning. Tyrone has consistently admonished me to not act like a whore. "If they don't wanna learn they ain't gonna learn. Leave 'em alone. When they wanna learn they'll come to you." Using my own metaphor of searching for doors, I have resolved to continue to identify doors that Tyrone might choose to open and enter. I cannot and will not do more than that. Nor will I back away. When Tyrone decides that the time is right, I will be there to facilitate his learning. In the meantime I show him respect and call him regularly to keep in touch. I worry a great deal about agency in my actions. I am a teacher and a researcher and everything I have done with Tyrone has so far been more advantageous to me than to him. While in the process of failing high school, he has co-authored a paper with me; he has assisted prospective teachers from Penn to better understand the teaching and learning of science in urban communities, and he has interviewed his peers and parents, thereby accessing data sources for our research that would otherwise be unattainable. If anyone is keeping score, I am way in the lead of who is receiving benefit from our relationship. It is a cruel paradox that someone as talented, smart, and thoughtful as Tyrone can be so marginalized by schooling. As science educators we must resolve such paradoxes. It is not enough to improve education for only the best students at City High.

Tyrone is intelligent, articulate and logical in the way he describes his goals and the raw deal he has been dealt at City High. Also he is optimistic when he describes his progress and the possibility of graduating in the near future. Tyrone's perceptions on his schooling at City High are encapsulated in the following excerpt from an interview with him.

> I don't like it because for the simple fact that what they teaching me is too easy. It's not the teachers there, or the students, or how it is. It's the work. 'Cause it's like, they give me some work to do. I'm finished in about 20 minutes. Each period is an hour and fifteen minutes long. So now I'm sitting there for about 45 minutes doing nothing. . . . The work they give me is too just too easy. (Tyrone, 05/99)

Tyrone's understanding of science is not strong, and yet his claim that the work is easy is typical of those made by his peers. Recently Tyrone undertook an investigation in which he immersed a one-cent coin into a solution of zinc ions. He knew already that the coin would "turn to silver," and based on a science activity from last year he predicted that it would "turn to gold" if he heated the silver coin in the flame of a Bunsen burner. Wearing his lab coat and glasses, Tyrone participated earnestly in the activity, and when he had confirmed his predictions he further claimed that when he came back the next day the coins would have reverted to their former state. His explanation was that the color would just rub off. A few days later when Tyrone came to school, I handed him the coins. His eyes glowed with pride. "Can I keep them?" I nodded affirmatively as he placed them in his pocket. I asked him about the color not rubbing off. He looked at them carefully and then affirmed that the colors had rubbed off last year. He returned the coins to his pocket and did not initiate further discussion about what had happened or why it might have happened. Tyrone was done with this activity and was ready to move on. He understood what had happened and could describe what he had done and how the coins had changed. However, he did not have concepts to explain what was happening at the microscopic/abstract level and had no apparent interest in building such models or a chemistry-like discourse.

Voiceover: Making connections between the students' primary discourses and the target secondary science discourses (The New London Group, 1996) has not been easy. There are at least two aspects to this. First, recognizing what students are good at, their funds of cultural knowledge, and figuring out how to scaffold lessons so that they can coparticipate in a process of building a new secondary science-like discourse would be a challenge at the best of times. However, since I had little rapport with my students and only a few of them allowed me to show my respect for them, it was difficult to promote productive verbal interaction that could lead to the negotiation of a constantly evolving shared discourse that became science-like.

During the past year I have had numerous conversations with Tyrone about my teaching and more particularly his experiences as a student and how to teach students like him. In a recent conversation he made a set of remarks about teaching students like him that were very salient to my teaching in *Opportunity*. He noted that:

> It's all in the teacher's attitude. 'Cause some teachers just see it like this. I'm the teacher. I got the authority. You gonna listen to what I say. Students, they don't like that no more. It not like it was back in the day when you would spank a student for not doing what they supposed to be doin. Nowadays you gotta treat a student just like you wanna be treated with a little respect. If not then that student ain't gonna give you no respect. I know like me. The way I see it—I respect all teachers if they respect me.

You don't respect me why should I respect you? Respect is a two way street from what I learned.

   Certain students, if they don't like a certain teacher then they won't do the teacher's work. They won't listen to the teacher. They'll ignore the teacher and there's nothin' the teacher can do about it. So . . . like . . . the more they try to teach them . . . the more they try to teach that . . . they really don't care for what the teachers talkin' about—even if they're interested in what the teachers talkin' about they don't like that teacher.

As soon as I heard Tyrone use the word respect I knew at once that I had failed to fully understand just how important respect and disrespect were in this culture. I had carried into this classroom my own *habitus* for teaching and in so doing had attempted to create a classroom that was like those middle-class, White classrooms in which I had been most successful. The failure of my *habitus* to meet the test of the community in which I was teaching exaggerated my lack of *Spielraum*. Rarely did I have room to maneuver and exercise pedagogical alternatives without deliberation. On the contrary, I frequently found myself coming out of action as I considered options reflectively. Frequently my *habitus* was in a state of breakdown, and the need to be reflective was a clear signal of my struggle to succeed. My inability to adapt my *habitus* to the unfolding circumstances inadvertently led me to demonstrate a fundamental and clearly visible disrespect for my students, what they could do, and their interests.

Voiceover: Without the respect of students it is difficult to see how they can construct me as teacher. Hence, in the order of things it seems an order of magnitude more important to build rapport and respect than to select particular activities in which students are to engage. This observation is consistent with the students' identification of the teacher they liked best. A graduate from City High, who had not yet gained a university degree or teaching credentials was the mathematics teacher in Opportunity. *Even though they rarely engaged in the mathematics and frequently quarreled in his class, the students consistently named him as their best teacher. He showed them respect and they reciprocated and were prepared to assign him the privilege of being their teacher, a privilege I was denied for most students.*

Using the Code of the Street to Run Up the Score

The biggest problem that arose during my teaching of science concerned the failure of the classroom community to follow the growth trajectory I had anticipated. Although I expected coparticipation to occur throughout a classroom community in which a shared language was negotiated and used by all and evolved to meet mutually shared goals, this did not occur. The building of a sense of community was thwarted by the students' sporadic attendance and by a lack of interest to engage in science on the part of those who actually were attending. Instead there was a struggle for control and, as a teacher, I had little power to establish the patterns of participation that I expected would support the learning of science.

Voiceover: Even though my efforts to introduce scientific terminology were done cautiously, most students regarded my language as difficult to penetrate and above them. They did not seem to be motivated to learn elementary terms that are essential to talking science and building deep understandings.

What was missing from the community? It is easy to respond to a question like that in terms of deficits and perhaps the question only invites deficit responses. At times I felt as if everything was lacking. The students seemed not to have the necessary conventions to handle scientific thinking and participation. It was as if they did not see science in their experiences. I wanted them to code switch, to leave behind the code of the street and employ patterns of interaction that might be described as science-like. For example, I wanted them to be curious in asking questions and seek alternative answers to questions. I also wanted them to be open minded with respect to the answers obtained by themselves and others, to be prepared to put their answers to empirical tests, and to examine the coherence of new understandings with their extant knowledge. Also, I wanted them to attempt to convince others of the viability of what they had learned and to speak and write about their growing knowledge such that others could learn from them. Few of my goals for them were fulfilled. Instead of participating as I wanted, many students seemed to earn the respect of their peers by winning public battles with me, by putting me down and thereby showing others in the class that they would not be dominated. My science classroom had its mores, but I did not know them well and was not in control of the rules. The students were operating according to the code of the street, and I did not know how to participate in a community that was evolving according to this code. Whereas I regarded the currency of this classroom community as knowledge, the students regarded the essential currency as respect.

Voiceover: Some of my approaches, such as monitoring, necessitated that I assign tasks and then circulate through the class to ensure students participated appropriately. My movement throughout the class was interpreted as "being in their faces," and I was consistently told to "back off, man." Being present to assist them was interpreted in terms of not trusting them to work without supervision, and the lack of trust was interpreted as lack of respect.

Throughout my teaching at City High I struggled to earn the respect of students and to build rapport. I was perplexed by their constant rejection of me and worked harder to build rapport based on my sincerity to teach them science and a commitment to not quit. I wanted to be courteous, fair in my dealings with them, respectful in my interactions with them, and informative in assisting them to benefit from a curriculum intended to make a difference to their lifeworlds. However, from the moment I arrived at school, I failed in my efforts to make headway. As I sat outside of the classroom waiting for Spiegel to arrive, I had only intermittent conversations with the students, and most of them interacted with one another using street language, much of it riddled with obscenities.[1] From time to time I requested students not to use obscene language in the school, but the use of obscenities in their conversations and rap songs was so pervasive as to be a torrent. For the most part my efforts to interact with students in the hallway were ignored, and attempts to control the use of obscene language were rejected.

---

1. City high has a policy of zero tolerance with respect to the use of obscene language, and teachers are expected to enforce the policy. In fact, few teachers make any effort to prevent the use of obscenities.

Voiceover: In the myriad interactions I have had with Tyrone, he has never once uttered an obscenity. He respects me too much to use such language. The continual use of obscene language by so many students in my presence and in the presence of teachers is a sign of disrespect for me, teachers in general, and the institution of schooling.

Acting alone I do not believe I can change the way these students from *Opportunity* interact with one another in the hallways. It will take a much larger set of forces to catalyze changes, and it will necessitate students accepting some responsibility for a different way of interacting while at school. I believe that two forces are operating. First, students interact with one another using a language to communicate and survive in their neighborhoods that maintains their identities and re-presents their differences and at the same time rejects standard English and all that it stands for. Second, the students earn respect from their peers by using language in my presence that they know I cannot fully understand and some that I cannot condone. I am powerless to prevent their use of their primary discourse, and even if I had the power to prevent its use I do not think I should do that. If I do nothing they accumulate respect from peers, and if I endeavor to prevent them from using obscenities, they earn even more respect by ignoring, confronting, or ridiculing me.

For these students the move from the hallways to the classroom is never something a teacher can take for granted. Many students prefer to remain in the hallways interacting with their peers until they are required to enter the classroom. Some teachers deal with this problem by locking the doors so that if they are not inside by the start of class they cannot enter at all. I do not believe in this practice, and students take advantage of my belief that they will come to class on time if they can only begin to value learning. Inadvertently my preferences play into the code because students can earn respect by moving to and from the classroom into the hallways. Even when students are in the classroom, some of them are enticed into the hallways by peers and the promise of getting to a bathroom to smoke a cigarette or to meet up with a boyfriend or girlfriend. Unless I am prepared to threaten the students with a punishment that will inconvenience them, such as suspension or detention, I am powerless to prevent them from "running up the score" as they please themselves about what they will and will not do. On any given day the most cooperation I can expect is from a handful of students. The majority is prepared to take advantage of my relative powerlessness and use myriad opportunities to gain the respect of their peers by "dissing" (showing disrespect for) their teacher.

Conclusions

Educational reform must always involve much more than exhortations to improve. Sitting on the side of Spiegel's classroom, it was very evident to me what was wrong and just as clear as to what needed to be done to correct the wrongs. As a teacher with a relatively long history of successfully teaching science it never occurred to me that what I had done elsewhere would not be successful with Spiegel's students. I was wrong on several accounts. The approaches that I considered most appropriate involved changing the nature of the curriculum so that students could employ what they can do, know, and are interested in. I believed in the importance of listening to the voices of the students and endeavored to do that. However, reading and thinking about reform are quite different from enacting reform. In this study I attained few of my goals as a teacher, and the main reason for my failure was an inability to see capital in the actions of the students. Despite my efforts to avoid deficit thinking and a stubborn resistance to lowering my expectations for student achievement and performance, what occurred as the curriculum was enacted was barely recognizable as science, and students were seldom stretched cognitively. Despite my efforts to the contrary I was unable to engage most students in higher cognitive thinking about science.

There were times when observers would come to my class and think they were seeing good science, however, this only reinforces the folly of making judgments from the side of a classroom. Some of my students, like Tyrone, are accomplished at getting the activities completed and achieving results such as forming brass on a copper coin. What is more of a challenge for most students is creating a science-like discourse in conjunction with such activities. What that takes, and what was missing from my classroom for the entire semester, is motivation to learn on the part of learners. Problems might be anticipated when a teacher has goals for students to learn science while the students are motivated to accomplish other goals, especially those related to attaining the respect of their peers. The negotiation of goals is an especially important stage in the enactment of a curriculum. In this instance I did not interact effectively with most students. My lack of cultural knowledge about my students was a decided disadvantage to being an effective teacher. I have much more to learn about how to be a teacher within a community in which respect is so important. I need to learn how to interact as a teacher for these students while not showing disrespect for them or requiring them to participate in ways that might be seen as earning the disrespect of their peers. At the same time it is necessary for me to earn respect while being consistent in my actions and beliefs about self. For example, I believe that successful learning communities are self-regulating, and I will not set up a classroom based on punishments and disincentives. Nor do I believe in establishing authority through shouting and physical aggression (involving the use of non-teaching assistants who are in the school to maintain a safe environment and to quell acts of physical aggression). I do not believe in suspending students for five days to teach them who is in charge of the school and would prefer to keep students in school rather than act in ways that return them to the streets. However, I do understand that the interests of those who aggressively resist efforts to maintain environments that are conducive to learning may have to be secondary to the interests of those of students who seek to learn. I prefer a community that is self-regulatory. In such a community the students could interact with me about the curricular goals, the nature of the activities to be undertaken at a given time, how to assess what has been learned and accomplished, and the rules for participation. Of course a precursor to any of these events occurring is that I must learn how to build and sustain a rapport with students that is constituted in mutual respect.

It would be advantageous if students came from elementary and middle schools with the discursive resources on which a scientific discourse could build. They do. But teachers like me are not good at recognizing the capital in what the students know and can do. It is evident that in the neighborhood city schools the resources that elementary and middle school students bring with them from their schooling and home environments does not fit as well with a science-like discourse as is the case with middle school students graduating from suburban schools around the country. We cannot continue to ignore the needs of high school students from working and unemployed classes and rationalize what is possible in terms of deficits emanating from experiences in urban elementary and middle schools and in the students' homes. Instead, it is time for educators to step up to the plate and spend the time needed to establish a community with its associated mores so that urban students can achieve science learning by coparticipating in safe and pleasurable learning environments.

The students in *Opportunity* would be a test for the best of teachers, and they are not typical of students from City High. Students are assigned to *Opportunity* because they have not been successful achievers or participants in high school, often due to an inability to settle down and accept their roles as learners. I opted to teach these students because of a sense that the group was abandoned. The rationale was strong. *Opportunity* exists so that others can learn. As repugnant as that is to me, I understand the rationale. However, in the past year City High has expanded from one *Opportunity* SLC to two, and just recently to three. This is an alarming trend, and it takes little imagination to envision an entire urban school as *Opportunity*. The trend must be reversed by teachers and students acting together to create learning communities in which science and other subjects can be taught and learned in a context of safe, caring, and enjoyable circumstances. Few teachers acting alone with a small set of students can establish and maintain the practices leading to productive learning. Heroic acts of particular teachers, as pleasing and inspirational as they might be, are not sufficient for the reform of urban schools. Nor is an exhortation for collaborative action. The doing of reform extends beyond written or spoken exhortations. From my perspective the initial step is to decide what type of community *we* intend to be and that cannot be decided in isolation from the students.

Whatever it takes, the first and most critical step is to decide what the community stands for, what is done in this community, and what happens when people interfere with others' efforts to attain the legitimate and negotiated goals of the community. Within that context it is possible to imagine science for all becoming a possibility at City High and other urban high schools. For my own part, prior to teaching physics in the spring of 2000, I am working with the principal of City High and teachers within *Opportunity* to establish community norms that highlight the significance of inquiry and problem solving.

My building of an appropriate *habitus* for teaching science in urban schools necessitated more than experience of teaching science in urban schools. Prolonged interactions with students such as Tyrone, in one-on-one situations, allowed me to understand how to interact with him in ways that were respectful and how to better understand his perspectives on life. My own culture, associated with being white, middle class, and Australian was so different from Tyrone's and his classmates that I was unable to build the rapport and respect that was essential for them to construct me as their teacher. No matter what I tried to teach, *Street Science* or *Chromatography*, the students were unlikely to cooperate and allow me to be their teacher. On the contrary, working one-on-one with students like Tyrone, interacting with students from *Opportunity* in the hallways for several months, and coteaching with numerous student teachers at City High enabled me to reengineer my *habitus*. My *habitus* changed to the extent that my interactions with students from *Opportunity* are now such that I feel more comfortable in predicting success when I teach them on future occasions. Not only did I build knowledge that could be represented in speaking and writing, but I also have adapted my *habitus* such that I can now interact differently and my *Spielraum* does seem to provide room to maneuver in the classes in which I am coteaching. One implication for teacher education is the significance of learning about culture and building the associated *habitus* and *Spielraum* by engaging in projects with students from diverse social and cultural communities.

What are the contexts for (re)learning to teach science, this time to relatively poor African American students in urban high schools? Starting out, I assumed that I knew how to teach, and I was surprised and embarrassed when my *habitus* for teaching broke down when put to the test (Roth & Tobin, 2001a). The issue for me was then to identify appropriate resources to support my learning of how to teach these students. Reflection on action was an obvious source for learning, and for many hours a day I searched through what I knew and read widely as I endeavored to create the knowledge that would lead me to success. Then it struck me, no amount of reading and re-conceptualizing was going to resolve the classroom problems I was encountering and which were exacerbated by my efforts to succeed. Not only that, being more deliberative in the classroom was creating uncertainty about what to do and creating feelings that I could not succeed. My attention then turned to the development of *habitus* and *Spielraum* and the resources to support the growth of each (Roth, 1998). Coteaching with Spiegel was one obvious resource as was coteaching with a doctoral student with experience in urban science teaching and with a student teacher who was assigned to teach science in *Opportunity* (Tobin, Seiler & Smith, 1999). However, the essential aspects of *habitus* that needed to be developed and the associated *Spielraum* that would enable me to anticipate in action necessitated regular one-on-one interactions with Tyrone so that I could learn how to read and appropriately respond to the signs of urban African American youth. Finally, my *habitus* was reconstructed to incorporate cultural knowledge grounded in an urban ethnography that I undertook. For more than a year I explored the neighborhoods of Philadelphia and built an identity of being in the streets and understanding the interactions between people with diverse social and cultural backgrounds. In the past year I have (re)learned teaching as praxis and at last have the confidence to teach science in ways that are likely to be more successful than my efforts of the spring and summer of 1999. Without wanting to sound too brash, I conclude this chapter with an exhortation to researchers, teacher educators and policy makers. Teaching science appropriately in urban settings extends far beyond selecting what is and is not an appropriate focus for the curriculum. Unless teachers are directed toward the resources for building viable *habitus* and appropriate *Spielraum* for urban participants, it is unlikely that we will see the improvement in urban science education that our citizenry yearns for and deserves. The issue of whether or not science education can be liberatory and emancipatory extends beyond the prescription of curricular foci and includes as central the issue of building a community constituted in rapport and mutual respect of all participants.

References

Anderson, E. (1990). *Streetwise: Race, class, and change in an urban community*. Chicago: University of Chicago Press.

Anderson, E. (1999). *Code of the street: Decency, violence, and the moral life of the inner city*. New York: Norton.

Barton, A. C. (1997). Liberatory science education: Weaving connections between feminist theory and science education. *Curriculum Inquiry, 27*, 141–163.

Barton, A. C. (1998). Reframing "science for all" through the politics of poverty. *Educational Policy, 12*, 525–541.

Bourdieu, P. (1992). *Language and symbolic power*. Cambridge, MA: Harvard University Press.

Lee, C. D. (1999). *Supporting the development of interpretive communities through metacognitive instructional conversations in culturally diverse classrooms*. Paper presented at the annual meeting of the American Educational Research Association, Montreal, Canada.

Mirón, L. F. & Lauria, M. (1998). Student voice as agency: Resistance and accommodation in inner-city schools. *Anthropology & Education Quarterly, 29*, 189–213.

National Commission on Excellence in Education. (1983). *A Nation at risk: The imperatives for educational reform*. Washington, DC: U.S. Department of Education.

National Research Council (1996). *National science education standards*. Washington, DC: National Academy Press.

New London Group (1996). A pedagogy of multiliteracies: Designing social futures. *Harvard Educational Review, 66*, 60–92.

Rodriguez, A. J. (1997). The dangerous discourse of invisibility: A critique of the National Research Council's national science education standards. *Journal of Research in Science Teaching, 34,* 19–37.

Roth, W.-M. (1998). Science teaching as knowledgeability: A case study of knowing and learning during coteaching. *Science Education, 82,* 357–377.

Roth, W.-M., & Boyd, N. (1999). Coteaching, as colearning, in practice. *Research in Science Education, 29,* 51–67.

Roth, W.-M., Lawless, D., & Masciotra, D. (2001). Spielraum and teaching. *Curriculum Inquiry, 31,* 183–208.

Roth, W.-M., Masciotra, D., & Boyd, N. (1999). Becoming-in-the-classroom: A case study of teacher development through coteaching. *Teaching and Teacher Education, 17,* 771–784.

Roth, W.-M., & Tobin, K. (2001a). *At the elbow of another: Learning to teach by coteaching.* New York: Peter Lang.

Roth, W-M., & Tobin, K. (2001b). Learning to teach science as praxis. *Teaching and Teacher Education, 17,* 741–762.

Rutherford, F. J., & Ahlgren, A. (1990). *Science for all Americans.* New York: Oxford University Press.

Tobin, K. (1987). The role of wait time in higher cognitive level learning. *Review of Educational Research, 57,* 69–95.

Tobin, K. (1999). The value to science education of teachers researching their own praxis. *Research in Science Education, 29,* 159–169.

Tobin, K. (2000). Becoming an urban science educator. *Research in Science Education, 30,* 89–106.

Tobin, K., Seiler, G., & Smith, M. W. (1999). Educating science teachers for the sociocultural diversity of urban schools. *Research in Science Education, 29,* 68–88.

Tobin, K., Seiler, G., & Walls, E. (1999). Reproduction of social class in the teaching and learning of science in urban high schools. *Research in Science Education, 29,* 171–187.

Toussaint, K. C. (1997). *Domination, power and racial stereotypes: Towards an alternative explanation of black male school resistance.* UMI Microform No. 9738009. Ann Arbor, MI: UMI Dissertation Services.

Whose Scientific Knowledge?
The Colonizer and the Colonized

Glen Aikenhead

At a 1982 science teachers' conference in Saskatoon, Canada, Jacques Désautels explained how conventional science teaching, claiming to transmit value-free knowledge to students, subliminally inculcates scientific and societal values. Like the Greek wooden horse during the siege of Troy, a science curriculum plays the role of a Trojan horse by concealing its values when teachers attempt to enculturate students into Western science. These values often take the form of an ideology called "scientism" (Ogawa, 1998; Smolicz & Nunan, 1975; Ziman, 1984). Nadeau and Désautels (1984) identified five ways in which this ideology surfaces in school science. First, there is a *naive realism*: scientific knowledge is the reflection of things as they actually are. Second, there is *blissful empiricism* according to which all scientific knowledge derives directly and exclusively from observation of phenomena. Third, there is *credulous experimentalism*, which holds that experimentation makes possible conclusive verification of hypotheses. Fourth, people committed to *blind idealism* believe that scientists are completely disinterested and objective beings in their professional work. Finally, those subscribing to *excessive rationalism* hold that the logic of science alone brings us gradually nearer the truth. Science teachers tend to harbor a strong allegiance to values associated with scientism, for instance, science is authoritarian, non-humanistic, objective, purely rational and empirical, universal, impersonal, socially sterile, and unencumbered by the vulgarity of human bias, dogma, judgements, or cultural values (Aikenhead, 1985; Brickhouse, 1990; Gallagher, 1991; Gaskell, 1992). Concealed in a Trojan-horse curriculum, scientism and other values penetrate students' minds when they learn to "think like a scientist" and take on other "habits of the mind"—goals emphasized in recent reform documents (AAAS, 1989; NRC, 1996). These new science curricula attempt to enculturate all students to the same value system.

Towards a Cross-Cultural Science Education

Enculturation is not a problem for a small minority of students whose worldviews resonate with the scientific worldview conveyed most frequently in school science (Cobern & Aikenhead, 1998). These "potential scientists" *want to* think like scientists (Costa, 1995). They embrace enculturation into Western science (Aikenhead, 1996; Hawkins & Pea, 1987). To them, there is no Trojan-horse curriculum.

But for the vast majority of students, attempts to enculturate them into Western science are experienced as assimilation into a foreign culture. These are the future citizens who will make strategic decisions for themselves and their society increasingly influenced by science and technology (Aikenhead, 1980; McGinn & Roth, 1999). Because they reject assimilation into the culture of Western science, they tend to become alienated from a major global influence in their lives. Alienation reduces their effectiveness at "legitimate peripheral participation" in community matters related to science and technology (Roth & McGinn, 1997).

The problem of alienation is more acute for Aboriginal students, whose worldviews, identities, and mother tongues create an even wider cultural gap between themselves and school science (AAAS, 1977; Cajete, 1986, 1999; Snively, 1990; Sutherland, 1998). For centuries, attempts to assimilate Aboriginal peoples into Euro-Canadian society (i.e., colonization) have had disastrous consequences (Battiste, 1986; Buckley, 1992; Deyhle & Swisher, 1997; MacIvor, 1995). Any further attempt to assimilate

Aboriginal students into Western science continues this colonization, and raises issues of social power and privilege in the science
classroom.

These issues formed the basis of a socio-cognitive model of
teaching and learning. Drawing upon the social cognitive work of
Delpit (1988), Lave (1988), and Wertsch (1991), O'Loughlin
(1992) persuasively claimed:

> To the extent that schooling negates the subjective, socioculturally con-
> stituted voices that students develop from their lived experience . . . and
> to the extent that teachers insist that dialogue can only occur on their
> terms, schooling becomes an instrument of power that serves to perpetu-
> ate the social class and racial inequities that are already inherent in so-
> ciety. (p. 816)

This model for equity science education is an alternative to the
conventional, uni-logical, assimilative, authoritative discourse that
transmits scientific knowledge and values to students. O'Loughlin
focused on "dialogical meaning making" in the context of social
power, thereby sharing the transformative goals of critical pedagogy (Freire, 1970):

> Dialogical meaning making occurs when the learner is influenced by the
> text, but is also allowed the space to play an active role in developing a
> personally constructed understanding of the author's or teacher's mes-
> sage through a process of dialogic interchange. (O'Loughlin, 1992, p.
> 813)

The discourse of instruction, O'Loughlin proposed, involves
more than the conventional literacy for comprehension (reading
the lines in science textbooks to infer comprehension, usually to
pass exams and acquire credentials). His discourse of instruction
is more than literacy for critical thinking (reading between the lines
to infer hidden assumptions, alternatives, and changes of meaning). For O'Loughlin one learns "to *participate* in the culture of
power, while simultaneously learning how to *reflect critically* on the
power relations of which they are a part" (p. 807, italics in the
original). His discourse of instruction is more like van der Plaat's
(1995) reading between the lines of privileged discourse to infer
what ontology has been culturally constructed by that discourse
and to understand that ontology in terms of its relationship to

one's own culturally determined ontology. This type of literacy is very much needed by many Aboriginal students (Cajete, 1999; MacIvor, 1995).

O'Loughlin's (1992) socio-cognitive model of meaning making addresses social power and privilege in the classroom, but it does not explicitly treat meaning making from a cultural perspective. A cultural perspective on science education is founded on such assumptions including the following. First, Western science is a cultural entity itself, a subculture of Euro-Canadian society. Second, people's cultural identities may be at odds with the culture of Western science. Third, science classrooms are subcultures of the school culture. Fourth, most students experience a change in culture when moving from their lifeworlds into the world of school science. Fifth, learning science is a cross-cultural event for these students (Aikenhead, 1996; Aikenhead & Jegede, 1999). These assumptions help to define a cultural approach for school science, one that tends to privilege science for all.

This approach to teaching and learning engages students in cultural negotiations in a context in which learning science is experienced as "coming to knowing," a phrase borrowed from Aboriginal educators (Ermine, 1998; Peat 1994). Coming to knowing is reflected in participatory learning: "If the living, experiencing being is an intimate participant in the activities of the world to which it belongs, then knowledge is a mode of participation" (Dewey, 1916, p. 393). The world in which most Aboriginal students participate is not a world of Western science but another world increasingly influenced by Western science and technology.

Coming to knowing engages Aboriginal students in their own cultural negotiations among the several sciences found within their school science, in which students become more aware of four aspects of their lifeworlds. First, students reflect on their own understanding of the physical and biological world. Second, students come to know the Aboriginal common sense understanding of their community. Third, they may encounter ways of knowing of another culture, including those of other First Nations peoples. Fourth, they are introduced to the norms, beliefs, values and conventions of Western science. This is known as "multi-science education" (Ogawa, 1995). Coming to knowing is also about developing cultural identity and self-esteem.

As mentioned above, a cultural approach to science education recognizes that learning Western science for most Aboriginal students is a cross-cultural event. Students move from their everyday cultures associated with home to the culture of Western science (Aikenhead, 1997; Phelan, Davidson, & Cao, 1991). These transitions, or border crossings (to use Giroux's [1992] metaphor), are smooth for "Potential Scientists," are manageable for other "Smart Kids," but are most often hazardous or impossible for everyone else (Costa, 1995). Success at learning the knowledge of nature of another culture for the purpose of coming to knowing depends, in part, on how smoothly one crosses cultural borders. Too often students (Aboriginal and non-Aboriginal alike) are left to manage border crossings on their own (Phelan et al., 1991). Most students require assistance from a teacher, similar to a tourist in a foreign land requiring the help of a tour guide. In short, a science teacher needs to play the role of a culture broker (Aikenhead, 1997).

Such a culture broker understands that Western science has its own culture, given that scientists generally work within an identifiable set of cultural attributes: "an ordered system of meanings and symbols, in terms of which social interaction takes place" (a definition by cultural anthropologist Geertz, 1973, p. 5). More specifically, the scientific community generally has its own language, beliefs, values, conventions, expectations, and technology. These attributes define a culture (Aikenhead, 1996). For Western science, these attributes are identified as "Western" because of the fact that the culture of Western science evolved within Euro-American cultural settings (Pickering, 1992; Rashed, 1997). The culture of Western science today exists within many nations, wherever Western science takes place.

A culture brokering science teacher makes border crossings explicit for Aboriginal students by acknowledging students' personal preconceptions and Aboriginal worldviews that have a purpose in students' everyday culture. A culture broker identifies the culture in which students' personal ideas are contextualized and then introduces another cultural context, for instance, the culture of

Western science, *in the context of* Aboriginal knowledge. At the same time, a culture broker must let students know what culture he/she is talking in at any given moment (e.g., Aboriginal science or Western science), because teachers can unconsciously (implicitly) switch between cultures, much to the confusion of many students. (Some specific strategies to accomplish this are described elsewhere [Aikenhead, 1997; Cajete, 1999; Jegede & Aikenhead, 1999].)

To facilitate students' border crossings, teachers and students both need to be flexible and playful and feel at ease in the less familiar culture (Lugones, 1987). This will be accomplished differently in different classrooms. As O'Loughlin (1992) argued, it has a lot to do with the social environment of the science classroom, the social interactions between a teacher and students, and the social interactions among students themselves. Thus, a teacher who engages in culture brokering should promote discourse (Driver, Asoko, Leach, Mortimer, & Scott, 1994) so students are provided with opportunities to engage in the following three activity types. First, students should have opportunities for talking within their own lifeworld cultural framework without sanctions for being "unscientific." Second, students should have opportunities for being immersed in either their everyday Aboriginal culture or the culture of Western science as students engage in some activity (e.g. problem solving or decision-making in an authentic or simulated event). Finally, students should know in which culture they participate at any given time.

Effective culture brokers substantiate and build on the validity of students' personally and culturally constructed ways of knowing (Pomeroy, 1994). Sometimes bridges can be built between cultures, other times ideas from one culture can be seen as fitting within the ideas from another culture. Whenever apparent conflict between cultures arises, it is dealt with openly and with respect. (Aikenhead and Jegede [1999] describe cultural conflict in terms of "collateral learning.")

For Aboriginal students, it will be helpful to deal with Western science's social, political, military, colonial, and economic roles in history. Smooth border crossings cannot occur if a student feels that he or she is associating with "the enemy" (Cobern, 1996). By

acknowledging Western science's historical roles in the colonization of Aboriginals on Turtle Island (North America), a teacher
can address Aboriginal students' conflicting feelings towards the
culture of Western science, thus making a student feel more at ease
with learning that culture without accepting its values and ideolo-
gies. In short, a culture brokering science teacher identifies the
colonizer and the colonized and teaches the science of each culture
(Snively & Corsiglia, 2001).

Cross-Cultural Science Education as Praxis

Allen and Crawley (1998), Cajete (1986), Kawagley (1995), MacIvor (1995) and Snively (1995) provided specific recommendations for teaching school science to Aboriginal students. Based on
these recommendations, a collaborative team of Saskatchewan
science teachers, university personnel, and people in the teachers'
local community have developed instructional strategies and units
of study to support teachers wishing to become culture brokers for
grade 6 to 11 Aboriginal students (Aikenhead, 2000). One product of this research and development activity is a set of cross-
cultural science and technology units (CCSTUs). The units bring
Western science into the students' world, rather than insisting that
students go into a scientist's world (the conventional way of
teaching science as assimilation).

A cross-cultural science and technology unit first creates an
Aboriginal framework into which Western science and technology
can be placed. This introductory content is drawn from appropriate Aboriginal knowledge and may take the form of practical action relevant to a community (e.g., listening to an elder, going on a
snowshoe hike, or assisting a local wild rice harvest). The choice
depends on the unit.

If the objective was to teach Western science's systems of the
body (e.g., the circulatory, nervous, and immune systems), we
might begin with the topic "Healing." In most Aboriginal cultures,
healing is conceptualized into four aspects: emotional, physical,
mental, and spiritual. Instruction establishes an Aboriginal view of

healing in the science class, appropriate for the age group and sensitive to the local culture.

The community's Aboriginal knowledge has a valid place in this curriculum. For instance, traditional ecological knowledge (TEK) can be combined with various fields of Western science (e.g., ecology, botany, biology, medicine, or horticulture) to give students an enriched understanding of nature in line with sustainable development (Snively & Corsiglia, 2001). Some students discover that they already possess some Aboriginal knowledge, while others learn it for the first time. Students' Aboriginal knowledge is given voice in the science classroom, in the dialogic sense of voice described by O'Loughlin (1992) as involving both the speaker and the listener in mutual respect. Thus, a CCSTU begins by validating "the ways of knowing students bring to school by grounding the curriculum in their voices and lives" (p. 814). A dialogic voice means that a teacher learns from students and people in the community. A teacher models successful border crossing with his/her students. In this context, students' Aboriginal identity has a legitimate place in classroom instruction. The discourse of power no longer resides with the teacher; power is more evenly shared.

The introduction to a CCSTU constitutes a framework for the whole unit. Throughout the unit, students return to this familiar framework as needed. The actual time to establish an Aboriginal framework could be as short as a 15-minute review or as long as a several-day experience.

Values are particularly salient in Aboriginal cultures (Cajete, 1999). The introductory framework to a CCSTU will identify values that elders expect students to learn. The Saskatoon Tribal Council, for instance, developed an informal academic program for school-aged children ("Super Saturday") which draws explicitly upon the values associated with the 15 tipi poles. Each Saturday is devoted to a different value. The value becomes one of the themes for university instructors to convey to the young people who visit them on Saturdays. In a school science unit "Healing," for example, a key value may be *harmony with nature*. This establishes the habit of identifying values that underlie Western science when that content is studied later in the CCSTU. When scientific values are made an explicit topic of discussion, they are clarified and critiqued, thus circumventing the indoctrination endemic to assimilative conventional science teaching. Students learn to identify vestiges of scientism in the text and verbal discourses of their everyday lives. The ontology of the Western colonizer (the mathematical idealization of the physical world) becomes more apparent, freeing students to appropriate Western knowledge and technique without embracing Western ways of valuing nature. (See Ogawa's [1996] four-eyed fish metaphor for a Japanese description of such an appropriation and Krugly-Smolska [1994] for other cultures.) The value of developing scientific knowledge is fundamentally different for the two cultures. While Western science values "revealing nature's mysteries" for the purpose of gaining knowledge for the sake of knowledge and material growth, Aboriginal science strives for living with nature's mysteries for the purpose of survival (Aikenhead, 1997; Simonelli, 1994; Snively & Corsiglia, 2001). Thus, each value system orients a student differently towards nature (Ermine, 1995).

Having established an Aboriginal framework and identified key values, the next step in a CCSTU is a border-crossing event in which teacher and students cross the cultural border into Western science, *consciously* switching values, language conventions, conceptualizations, assumptions about nature, and ways of knowing. As a culture broker, the teacher clearly identifies the border to be crossed, guides students across that border, and helps students negotiate cultural conflicts that might arise. Because values are very important to Aboriginal communities, a teacher identifies a key value that underlies the Western science in the unit. For instance, in the unit "Healing," one value that underlies the science of body systems is "power over nature." The pharmaceutical industry is a case in point.

One feature that often emerges from comparing Aboriginal and Western science is the recognition that Western science can powerfully clarify one small aspect of Aboriginal science. For instance, in the "Healing" unit, Western medicine deals predominantly with the physical aspects of healing, and so Western science is seen as informing a small slice of the Aboriginal framework (with its four aspects of healing). When values are made explicit, students are only expected to recognize those values, not to adopt them for their own. The foreignness of Western science begins to feel less threatening. Social power and privilege in the classroom increases for students who sense genuine respect for their Aboriginal values.

As various topics in Western science are studied within the unit, it will make sense to include more Aboriginal content (more than in the introduction). This is easy to do because the unit already has a framework for that content. The Aboriginal content is not just tacked on for the sake of creating interest. It frames the unit in a way that nurtures the enculturation of Aboriginal students into *their community's* culture (Casebolt, 1972). This discourse of Aboriginal knowledge is very different from the discourse of Western science. Both have a function in the classroom. Students share their coming to knowing with their teacher in a dialogic manner. Students bring their community's knowledge and values into the classroom. New power relationships replace the conventional colonizer/colonized hierarchy.

During any lesson within a CCSTU, students are able to state which culture they are speaking in (Aboriginal or Western science). Culture brokering teachers can make this explicit, for example, by using two different black boards—one for Aboriginal science, another for Western science. One is used to record ideas expressed in the discourse of the community's Aboriginal knowledge, whereas the other board is used to express the culture of Western science. By switching from one board to the other (cultural border crossing), students switch language conventions, conceptualizations, values, assumptions about nature, and ways of knowing. It is up to the teacher to assess the quality of students' learning associated with both boards; both have a place in the assessment. A concrete approach like this helps students gain access to Western science without losing sight of their cultural identity. In fact their cultural identity is cultivated by the classroom's emphasis on coming to knowing.

Nelson-Barber and colleagues (1996) have mapped out the assessment of student achievement within a cross-cultural science classroom. They offer guidance and specific recommendations for developing a culturally responsive assessment system, beginning with the recommendation to treat linguistic and cultural diversity as strengths. An example from the Navajo (Diné) Nation demonstrated the fruitfulness of portfolio assessment. Portfolios were shown to promote student autonomy and reflected the *context* of learning, not just the process and product of learning. The international recognition of the efficacy of student self-assessment (Black & Atkin, 1996) lends credence to negotiating with Aboriginal students how school science will be assessed. Without such a negotiation, the balance of social power and privilege reverts back to the colonizer-colonized hierarchy.

In summary, culturally sensitive CCSTUs help Aboriginal students feel that their school science courses are a natural part of their lives. CCSTUs give students access to Western science and technology without requiring them to change their own cultural identity. They are not expected to adopt the worldview endemic to Western science. However, for those students who have a gift for Western science, a CCSTU lays the foundation for further education in science and engineering.

In either case, cross-cultural science and technology units represent one form of *science education as/for social action*. The units encourage a change in the power relationships between a teacher and her Aboriginal students in ways that promote mutual respect, coming to knowing, and the ethic of survival for humankind. As a result, teachers and students become better critical social actors in a Canadian society enriched by cultural differences but challenged by risks to human survival. Whose scientific knowledge will be taught in school science? The cultural capital of Aboriginal peoples can effectively contribute to ameliorating the colonizer/colonized hierarchy in science education to the benefit of both groups.

References

American Association for the Advancement of Science (AAAS). (1977). *Native Americans in science.* Washington, DC: Author.
American Association for the Advancement of Science (AAAS). (1989). *Project 2061: Science for all Americans.* Washington, DC: Author.

Aikenhead, G.S. (1980). *Science in social issues: Implications for teaching*. Ottawa, Ontario: Science Council of Canada.

Aikenhead, G. S. (1985). Collective decision making in the social context of science. *Science Education, 69*, 453–475.

Aikenhead, G. S. (1996). Science education: Border crossing into the subculture of science. *Studies in Science Education, 27*, 1–52.

Aikenhead, G. S. (1997). Toward a First Nations cross-cultural science and technology curriculum. *Science Education, 81*, 217–238.

Aikenhead, G. S. (2000). "Rekindling traditions: Cross-cultural science & technology units." Available at http://capes.usask.ca/ccstu.

Aikenhead, G. S., & Jegede, O. J. (1999). Cross-cultural science education: A cognitive explanation of a cultural phenomenon. *Journal of Research in Science Teaching, 36*, 269–287.

Allen, J. A., & Crawley, F. E. (1998). Voices from the bridge: Worldview conflicts of Kickapoo students of science. *Journal of Research in Science Teaching, 35*, 111–132.

Battiste, M. (1986). Micmac literacy and cognitive assimilation. In J. Barman, Y. Herbert, & D. McCaskell (Eds.), *Indian education in Canada, Vol. 1: The legacy* (pp. 23–44). Vancouver: University of British Columbia Press.

Black, P., & Aiken, J. M. (1996). *Changing the subject: Innovations in science, mathematics and technology education*. London: Routledge.

Brickhouse, N. W. (1990). Teachers' beliefs about the nature of science and their relationship to classroom practice. *Journal of Teacher Education, 41*(1), 52–62.

Buckley, H. (1992). *From wooden ploughs to welfare: Why Indian policy failed in the prairie provinces*. Montreal, Canada: McGill-Queens University Press.

Cajete, G. A. (1986). Science: A Native American perspective. Unpublished doctoral dissertation, International College, Los Angeles.

Cajete, G. A. (1999). *Igniting the sparkle: An Indigenous science education model*. Skyand, NC: Kivaki Press.

Casebolt, R. L. (1972). Learning and education at Zuni: A plan for developing culturally relevant education. Unpublished doctoral dissertation, University of Northern Colorado, Boulder.

Cobern, W. W. (1996). Worldview theory and conceptual change in science education. *Science Education, 80,* 579–610.

Cobern, W. W., & Aikenhead, G. S. (1998). Cultural aspects of learning science. In B. J. Fraser & K. G. Tobin (Eds.), *International handbook of science education* (pp. 39–52). Dordrecht, Netherlands: Kluwer Academic Publishers.

Costa, V. B. (1995). When science is "another world": Relationships between worlds of family, friends, school, and science. *Science Education, 79,* 313–333.

Delpit, L. (1988). The silenced dialogue: Power and pedagogy in educating other people's children. *Harvard Educational Review, 58,* 280–298.

Dewey, J. (1916). *Democracy and education: An introduction to the philosophy of education.* New York: Macmillan.

Deyhle, D., & Swisher, K. (1997). Research in American Indian and Alaska Native education: From assimilation to self-determination. *Review of Research in Education, 22,* 113–194.

Driver, R., Asoko, H., Leach, J., Mortimer, E., & Scott, P. (1994). Constructing scientific knowledge in the classroom. *Educational Researcher, 23*(7), 5–12.

Ermine, W. J. (1995). Aboriginal epistemology. In M. Battiste & J. Barman (Eds.), *First Nations education in Canada: The circle unfolds* (pp. 101–112). Vancouver: University of British Columbia Press.

Ermine, W. (1998). Pedagogy from the ethos: An interview with Elder Ermine on language. In L. A. Stiffarm (Ed.), *As we see ... Aboriginal pedagogy* (pp. 9–28). Saskatoon, Canada: University of Saskatchewan Extension Press.

Freire. P. (1970). *Pedagogy of the oppressed.* New York: Herder & Herder.

Gallagher, J. J. (1991). Prospective and practicing secondary school science teachers' knowledge and beliefs about the philosophy of science. *Science Education, 75,* 121–133.

Gaskell, P. J. (1992). Authentic science and school science. *International Journal of Science Education, 14,* 265–272.

Geertz, C. (1973). *The interpretation of culture.* New York: Basic Books.

Giroux, H. (1992). *Border crossings: Cultural workers and the politics of education.* New York: Routledge.

Hawkins, J., & Pea, R. D. (1987). Tools for bridging the cultures of everyday and scientific thinking. *Journal of Research in Science Teaching, 24,* 291–307.

Jegede, O. J., & Aikenhead, G. S. (1999). Transcending cultural borders: Implications for science teaching. *Research in Science and Technology Education, 17,* 45–66.

Kawagley, O. (1995). *A Yupiaq worldview.* Prospect Heights, IL: Waveland Press.

Krugly-Smolska, E. (1994). An examination of some difficulties in integrating western science into societies with an indigenous scientific tradition. *Interchange, 25,* 325–334.

Lave, J. (1988). *Cognition in practice: Mind, mathematics and culture in everyday life.* Cambridge, England: Cambridge University Press.

Lugones, M. (1987). Playfulness, "world"-travelling, and loving perception. *Hypatia, 2*(2), 3–19.

MacIvor, M. (1995). Redefining science education for Aboriginal students. In M. Battiste & J. Barman (Eds.), *First Nations education in Canada: The circle unfolds* (pp. 73–98). Vancouver: University of British Columbia Press.

McGinn, M. K., & Roth, W.-M. (1999). Preparing students for competent scientific practice: Implication of recent research in science and technology studies. *Educational Researcher, 28*(3), 14–24.

Nadeau, R., & Désautels, J. (1984). *Epistemology and the teaching of science.* Ottawa, Canada: Science Council of Canada.

Nelson-Barber, S., Trumbull, E. & Shaw, J. M. (1996, August). *Sociocultural competency in mathematics and science pedagogy: A focus on assessment.* A paper presented to the 8th Symposium of the International Organization for Science and Technology Education, Edmonton, Canada.

National Research Council (NRC). (1996). *National science education standards.* Washington, DC: National Academy Press.

Ogawa, M. (1995). Science education in a multi-science perspective. *Science Education, 79*, 583–593.

Ogawa, M. (1996). Four-eyed fish: The ideal for non-western graduates of western science education graduate programs. *Science Education, 80*, 107–110.

Ogawa. M. (1998). Under the noble flag of 'developing scientific and technological literacy.' *Studies in Science Education, 31*, 102–111.

O'Loughlin, M. (1992). Rethinking science education: Beyond Piagetian constructivism toward a sociocultural model of teaching and learning. *Journal of Research in Science Teaching, 29*, 791–820.

Peat, D. (1994). *Lighting the seventh fire.* New York: Carol Publishing Group.

Phelan, P., Davidson, A., & Cao, H. (1991). Students' multiple worlds: Negotiating the boundaries of family, peer, and school cultures. *Anthropology and Education Quarterly, 22*, 224–250.

Pickering, A. (Ed.). (1992). *Science as practice and culture.* Chicago: University of Chicago Press.

Pomeroy, D. (1994). Science education and cultural diversity: Mapping the field. *Studies in Science Education, 24*, 49–73.

Rashed, R. (1997). Science as a western phenomenon. In H. Selin (Ed.), *Encyclopaedia of the history of science, technology, and medicine in non-western cultures* (pp. 884–890). Boston: Kluwer Academic Publishers.

Roth, W.-M., & McGinn, M. K. (1997). Deinstitutionalizing school science: Implications of a strong view of situated cognition. *Research in Science Education, 27*, 497–513.

Simonelli, R. (1994). Sustainable science: A look at science through historic eyes and through the eyes of indigenous peoples. *Bulletin of Science, Technology & Society, 14*, 1–12.

Smolicz, J. J., & Nunan, E. E. (1975). The philosophical and sociological foundations of science education: The demythologizing of school science. *Studies in Science Education, 2*, 101–143.

Snively, G. (1990). Traditional Native Indian beliefs, cultural values, and science instruction. *Canadian Journal of Native Education, 17*, 44–59.

Snively, G. (1995). Bridging traditional science and western science in the multicultural classroom. In G. Snively & A. MacKinnon (Eds.), *Thinking globally about mathematics and science education* (pp. 1–24). Vancouver: Centre for the Study of Curriculum & Instruction, University of British Columbia.

Snively, G., & Corsiglia, J. (2001). Discovering indigenous science: Implications for science education. *Science Education, 85,* 5–34.

Sutherland, D. L. (1998). Aboriginal students' perception of the nature of science: The influence of culture, language and gender. Unpublished doctoral dissertation, University of Nottingham, Nottingham, UK.

van der Plaat, M. (1995). Beyond technique: Issues in evaluating for empowerment. *Evaluation, 1,* 81–96.

Wertsch, J. V. (1991). *Voices of the mind: A sociocultural approach to mediated action.* Cambridge, MA: Harvard University Press.

Ziman, J. (1984). *An introduction to science studies: The philosophical and social aspects of science and technology.* Cambridge, England: Cambridge University Press.

8

Reconstructing the Harsh World: Science with/for Social Action

Angela Calabrese Barton & Margery D. Osborne

The idea of science with/for social action raises the following questions for us as science educators working with minority and "at-risk" children: What is social action in science? In education? Where is it located and whose purposes does it serve? Where do children and teacher "fit" into science and science education with/for social action? In this paper we explore these questions by examining the concept of "homeplace," a construct brought to our science classes by children, which recasts the science as something transformative. Arguing from our position as feminists committed to social change, we demonstrate that critical articulations, acts of "remembrance," are fundamental to science and education with/for social action. These articulations concern both the sociohistorical lives of the children and teacher involved (and this includes the experiences, values and ideas they bring to the science classroom)—their "homeplaces"—as well as how these sociohistorical lives get positioned with and against science—the "harsh world."

To make this argument we draw from feminism, and in particular, ecofeminist philosophy. We develop the position that feminism in science education is about constructing places in which the enterprise of science can be rethought and "science" can be placed in a position (its proper position) as a tool for enacting societal change for the better. This in turn means a reconstruction of both "homeplace" (the child's situation) and the "harsh world," the science itself. In this sense the "knowing and doing" of science in schools is the active intersection between the lifeworlds of children, and the external worlds they come to study. Science education as social action is about pursuing those intentions and intersections and as such is about reconfiguring power relations for we assert that any social action is fundamentally tied to real and ideological shifts in power relations.

Such a line of thought draws formatively from articulations of Black (hooks, 1990; Collins, 1990; 1998) and Third World feminist writings (Minh-ha, 1989; Mohanty, 1991; Sandoval, 1991, 1995) as well as ecofeminist philosophy (Gaard & Murphy, 1998; Haraway, 1991, 1992; Merchant, 1990). Ecofeminism is a *practical* movement for social change arising out of the struggle of women to sustain themselves, their families, and their communities. These struggles occur in the context of the concrete, the places that people actually live within, and occur against both the abstract images of patriarchy, multinational corporations, and global capitalism and their concrete manifestations in peoples' lives. Donna Haraway (1991, 1992) in particular makes the point (in writing about the image of the "cyborg") that we cannot separate nature from culture and that culture is constructed within the context of our daily technologies and our immediate life setting. Indeed we must re-imagine nature to include the urban landscape where we live. This is, in turn, a constructed landscape, one we create/can recreate in the interplay between science (technology), nature, and ourselves and our needs. In this sense science becomes an empowering tool used to reconstruct an alienating environment so that we can incorporate it into our vision of "homeplace."

We argue that central to a construct of science education for social change is giving children the tools and power to do this re-creation. This is a redefinition of representations of traditional science for it resituates the "expert" in the child and redefines the elitist, abstract and distancing knowledge of the "expert" as ordinary and human, connected to lives, living and reflecting a sense of personhood. When we say this we think of Evelyn Fox Keller's description of Barbara McClintock's science as humanist in the sense that it reflected her empathy with the plants she studied and her own spirituality. Her science was a *manifestation of herself* in complex ways, which included her procedures in doing science, her interpretations of her observations and her final, ground-breaking articulations of genetic theory. They were revolutionary because they were *different*, and they were different because McClintock was different. The children we work with are also *different*, and their science is too and should be paid attention to just because of that. This different-ness informs us about what science in general could be about and, more specifically, what science for social action is.

Homeplace and Science for Social Change

Historically, African-American people believed that the construction of a homeplace, however fragile and tenuous (the slave hut, the wooden shack), had a radical political dimension. Despite the brutal reality of racial apartheid, of domination, one's homeplace was the one site where one could freely confront the issue of humanization, where one could resist. Black women resisted by making homes where all black people could strive to be subjects, not objects, where we could be affirmed in our minds and hearts despite poverty, hardship, and deprivation, where we could restore to ourselves the dignity denied us on the outside in the public world. (hooks, 1981, p. 41)

How can concepts of homeplace and the harsh world inform our understanding of science education for social change? We believe that a summative quality of science and science education for social change has to do with a radical political dimension which follows from understandings of personal lives and life situations contained within the acts of remembrance we develop here. We use these acts to begin to answer the questions we posed earlier (What is social action in science? In education? Where is it located and whose purposes does it serve? Where do children and teacher "fit" into science and science education with/for social action?) and to examine the concept of "homeplace" as a construct brought to our science classes by children which recasts the science as something transformative.

In the following, we tell a story from Angie Calabrese Barton's research. The story is about Darkside, a sixteen-year-old self-labeled Black Cuban American, who lives with his family in a homeless shelter in New York City. Darkside has lived in shelters for over three years and most recently at the shelter where we work for nine months. He lives with his mother and father and younger brothers. He plans to graduate from high school, to go to college, and to make something out of his life, although he is not sure what exactly it is he wants to do. He has worked hard to stay out of gang life—one older brother died violently and another is incarcerated. However, his efforts to stay away from gangs has made schooling difficult for within the social structuring of schools such as Darkside's gangs are both the source of conflict and safety. In early January he was chased by several teens, was tripped and broke his leg. Because he was in a full-leg cast for eight weeks and required several additional weeks to rebuild leg strength, he refused to attend school for his own safety.

The shelter where Darkside lives is in a poor, run-down area. There are over 200 families housed there for periods of time ranging from three to fifteen months. This particular shelter stands out from other shelters in the New York City system because of the network of services it provides for the children and parents. Although all shelters are required to provide families with social services, this shelter provides a myriad of educational programs. These programs include GED courses for teenagers and adults, after-school homework and tutoring for children ages 6–18, recreational activities such as basketball teams, and daycare and early childhood education for children under the age of six.[1]

However, many activities and regulations at the shelter are strikingly similar to the other shelters in the city. For example, all shelter residents are required to sign in and out and are held to strict curfews: 9 p.m. for children under the age of 12, 10 p.m. for children under the age of 18 and 11 p.m. for adults. Children and youth under 18 are not allowed on shelter property without an

---

1. General educational diplomas (GED) provide individuals who have not graduated from a regular high school program with an equivalent certificate and the opportunity to continue tertiary studies. Recipients must successfully complete and pass the General Education Diploma exam in order to receive such a certificate.

adult guardian. So, for example, if a parent is at work or elsewhere, children cannot go home. Visitors, including immediate family members such as brothers, sisters, parents, and grandparents, as well as friends or daycare providers not residing in the shelter are not allowed. Finally, the shelter is surrounded by bar fences, giving it a prison-like quality. In fact, the three things about living at the shelter that Darkside likes least are the security guards, the curfews, and the prison-like feel.

Throughout the entire school year that he lived at the shelter, Darkside was active in an action research project there. The action research project was something that my (Angie's) doctoral students, a post doc and I initiated with the teens at the shelter. Even though we initiated the project (we had the money to do so) the goals and purposes of the action research after our initial meetings were strictly up to the teens. We began by talking with them about their concerns and tried to find ways as a community to productively address those. The children rapped, made murals and role-played, expressing their interests and concerns in multiple ways. The teens talked about gangs, school, work, family, personal relationships, and child rearing, among other things. One of the concerns to emerge through these activities was about the lot across the street from the shelter. This particular lot was abandoned, full of litter including such items as ripped open garbage bags, feces, broken bottles, and crack vials. A partially destroyed metal link fence surrounded the lot with sharp fragments protruding in several places from a "high-speed police chase." The damaged fence was both an eyesore and unsafe. The teens decided to make this lot a focus of activity.

We worked with the teens to document the qualities of the lot, including its size, shape, and positive and negative attributes and used this information to generate a reasonable list of things we could do to make the lot something that they would enjoy. The teens generated many ideas including, a garden, a playground, a stage, a basketball court and an apartment building. Over the next two months, from these ideas, the teens did research to figure out what they might need to accomplish their options, built 2-D and 3-D models, and debated their ideas. They settled on a community garden with benches because it would be beautiful, would not be too expensive, and, although it required upkeep, that would be minimal, except during the summer months.

Darkside was actively involved in the entire process to transform the abandoned lot into a community garden. He attended the biweekly meetings faithfully. Although he was often quiet at these meetings, when he spoke, he did so with passion and authority, usually in an attempt to focus the other teens or to provide new ideas.

From the very beginning, Darkside viewed his work in the action research project as important:

> You want to change the environment and make a difference. That is what we are trying to do to this lot over here. (Darkside, March 9, 1999)
>
> It is important to do things for your community, to make it a better place. That's what we are doing with the lot. Beautify it. Help people want to be there, to spend time there. We need to help our community. (Darkside, March 9, 1999)

This connects for Darkside to images of "doing science" and the scientific community although only within his local setting in the Bronx. In fact, Darkside even argued with his peers to convince them that this kind of work makes them part of the scientific community:

Darkside: I think that when we are making the lot into a garden, then we are part of the scientific community.

Angie: Why?

Darkside: Because we using science to make the community better. We are part of science. We contributing.

Angie: Goldberg, do you agree or disagree with Darkside?

Goldberg: Fifty fifty.

Angie: Why do you say, fifty fifty?

Goldberg: Because, we not really part of science. We might be doing something for the community, but that doesn't mean we are part of the scientific community.

Darkside: Yeah, but if we use science, then we are part of the scientific community.

Goldberg: No, but not the scientific community.

Darkside: Mr. Goldberg, I'm saying that we know enough science. We use what we know about the environment and plants and pollution to transform the lot. We are doing this for the com-

munity. No one is telling us what to do or how to do the garden. So, we are part of the scientific community.
Goldberg: Yeah, yeah, yeah.

One of Darkside's major reasons for helping to transform the
abandoned lot into a garden is that he wants to be proud of his
community. He wants people to walk by, to stop and sit down,
and to enjoy what the garden has to offer. He feels he is mostly
doing science because he is testing the soil, planting vegetables and
flowers, cleaning up the pollution, recycling, and doing something
positive for his community. He feels like he has learned some science but only science that will contribute to his local community, a
place he is proud of and hopes to be remembered by.

Homeplace: Transforming the Harsh World

We use this story of Darkside and a description of his life at the
shelter and his views of science for two reasons: his vision of science has shaped what he does after school at the shelter and his
relationships with his peers. His vision is fundamentally an enactment of homeplace, remembrance and social action. So if we
look closely at the story of Darkside, what is homeplace and what
is science for social action, and how do the two relate? In our attempt to answer this question, we first begin with a discussion of
homeplace. We then move to a discussion of Darkside's construction of homeplace.

In one of her earlier pieces, bell hooks talks about the black
family, and in particular the black mother and grandmother. The
black family returns home from the "harsh world" to create a safe
space of care and nurture in the face of racist oppression and
sexist domination, and that reality, alone, made the home, regardless how fragile or tenuous, have a radical political dimension
(hooks, 1984). hooks refers to this nurturing space away from the
harsh world as *homeplace.*

In Black Feminist thought, homeplace has several important
qualities. First, homeplace anchors: Homeplace is a safe community. Such a community is built on difference and solidarity and is
a place where people who may or may not be "relatives" come to

know and rely on each other. This vision of community is important, especially in terms of how we think about young people. It is
important because it acknowledges that relationships between
parents and children go beyond children as property or posses-
sions of parents, to children and parents as members of a larger
caring community of on-going relationship, friendship and sites for
change. Understanding this also allows us to recognize the roots
of a shared history and a common anguish among community
members. It is in the community where a shared history and anguish is taken broadly to mean a common and articulated understanding and set of experiences around issues of oppression and
liberation and a critique of those experiences. This common history and anguish join members together in solidarity in a struggle
to be heard and treated fairly outside the homeplace.

Even though homeplace as community draws strength from a
common history and anguish, homeplace as community teaches—
and requires—participants to relate to a wide variety of people
and backgrounds. This idea of difference is important because it
asserts "unassimilated otherness" or representation, and "voice"
to members of smaller groupings. This, in turn, celebrates distinctions and characteristics of such different smaller groupings as one
way to motivate change among all larger-group members (Young,
1990, p. 319). By deconstructing the ideal of homogeneity within
community, it allows for a transformation of all members through
the recognition of difference. From this perspective communities
form "not through the negation of the given but rather as making
something good from the many elements of the given" (p. 317). In
this sense, community—and homeplace—is also built through a
politics of difference. And, the tension between common history
and difference keeps the community balanced, self-critical, and
responsive to its members.

Homeplace is also marked by several *practices* (for a summary,
see Table 1). For example, homeplace is a context where all members are affirmed through the act of *remembrance*. By remembrance
we mean a critical and creative articulation and reflection on the
homeplace, its members and their experiences. To us, this remembrance speaks to a painful history of subjugation. We think what
is important here is the role that remembrance plays in the pur-

Table 1: Home place

Qualities	- Homeplace is a safe community built from difference and solidarity
Practices	- Remembrance - Subversion - Community-based problem solving an decision making - Solidarity in action to solve identi-fied problems - Action to transform the harsh world
Knowledge and Values	- Respect for difference - Respect for life experiences - Experience based - Cultural and political

pose and the goals of the family or the community. Remembrance takes into account the historical situatedness of the articulation of how we come to know the world (in both what we know and how we know it). It also takes into account how the articulation of the historical brings with it a radical political dimension because it calls into question connections between position, power, and knowledge. In this sense, homeplace as the practice of remembrance also embraces acts of subversion: The very act of articulation, in remembrance, politicizes experience and the meaning of experience and opens up spaces for critique and revision of those experiences and the world which helped shape them.

Homeplace also carries with it its own orientation to knowledge and values. Homeplace challenges static representations of scientific knowledge or any kind of knowledge because the production of knowledge is connected to the social uses of and need for knowledge. In building science from homeplace this means starting from a belief that the *knowing* and the *doing* of science are historically, socially and politically situated processes influenced by external needs, and that scientific knowledge is also shaped through internal channels. Science is an outgrowth of those who create it, even when drawing from work historically constructed within the discipline. Neither scientific knowledge nor the constructor of that knowledge can be defined separately from the other—each requires the other. Perhaps more important, though, is the political dimension to knowledge construction. For us this means two things. First, because knowledge is experientially based, it is a representation of one's reality and can be used as lens to understand, critique and revision the realities of the harsh world, be it formal science or larger social contexts. Second, knowledge is always used for something, to understand and influence—to change—one's physical or existential reality.

Together, these ideas about community, practices such as remembrance and subversion, and knowledge and values embracing cultural and political standpoint underscore the cross-cutting theme that homeplace is ultimately about transforming the harsh world. Everything that happens in and about homeplace is centrally connected to the ideal of a fair and equitable world. This is why we refer to the idea of homeplace as radical and political. Homeplace exists through the efforts of those marginalized to create the physical, emotional, and intellectual space to understand, critique and recreate institutional and social practices based in the discourses of domination and control.

Discussion: Connecting Darkside with Homeplace, Ecofeminism, and Differential Consciousness

Darkside constructs a community among his peers in the transformation of the lot to the garden. He pushes them to consider how their visions for the garden and their subsequent actions are critical to revitalizing the South Bronx. He pushes them to consider how they might use their own constructions of science and self to sustain their community and their actions while also transforming the larger communities:

1. Physical transformation: Darkside works to transform the abandoned lot into a garden where people in the area can take respite.

2. Social: Darkside works to show the rest of the world (especially those who see the South Bronx as an ugly and scary place) that the South Bronx is beautiful and a place where its current residents want to live and are proud to live.

3. Scientific: Darkside works to convince his peers that they are already a part of the scientific community when they use science to enact change in their community. What is crucially important is that the teens use and produce the science of their choosing.

Embedded within his transformative acts are acts of remembrance: Informing Darkside's actions are his critical and creative articulation and reflection on his local community and the experiences, needs, and concerns of those within his community. Earlier, we described the importance of the radical political dimension that remembrance plays in the purpose and the goals of the family, or the community. Darkside's intentions, actions, and descriptions of his actions call into question connections between position, power, and knowledge. In this sense, homeplace—or science with/in the community—is constructed and used by Darkside. Here, Darkside's knowledge is experientially based; it is a representation of his reality and he uses his experience as lens to understand, critique and transform the realities of his local community. In the process, he also transforms science and his place in science.

In the essay "U.S. Third World Feminism" Chéla Sandoval (1995) offers an alternative "topography" (rather than a "typology") of feminisms which maps on to traditional feminist typologies (e.g., liberal, radical, cultural, socialist, etc.) and extends them in interesting ways. As well as providing a description of categories of feminism relevant to Third World women her topography makes clear that forms of oppositional consciousness are only possible as an intersection between the multiple struggles of the marginalized against gender domination, in the context of race, class, and cultural hierarchies. She suggests five categories of oppositional consciousness. The first is "equal rights," in which a group argues that their differences are only superficial and could be shed with equal access to resources or accommodated if understood by those in power. Second, "revolutionary," in which a group articulates their differences from those in power and claim these as central and valuable. The members of the group then call for a social transformation that will accommodate and legitimate those differences. Sandoval describes a third category, "supremacism," in which "not only do the oppressed claim their differences, but they also assert that those very differences have provided them access to a superior evolutionary level than those currently in power" and thus justify their leadership over the powerful. Next, "separatism," in which the oppressed claim their differences but do not aim for integration, transformation or leadership, but rather a "form of political resistance...organized to protect and nurture the differences that define it through complete separation from the dominant social order" (p. 415).

The final category, differential consciousness, is the most interesting to us as we try to understand the concept of homeplace and children like Darkside and their relationship to science. According to Sandoval (1991), differential consciousness "operates like the clutch of an automobile: the mechanism that permits the driver to select, engage gears in a system for the transmission of power" (p. 13). She sees it as operating through the other categories by changing each category's emphasis from a fixed set of positions, ideas, and analyses to a fluid set of tools, tactics, and approaches. These latter approaches are to be used when the situation calls for them—particularly in cobbling together coalitions that enable mobilization of the means to resist external oppressive forces. Differential consciousness allows an actor to focus on power hierarchies and enables subversion of and resistance to power relations. Rather than polarizing by focusing on differences, differential consciousness allows us to see where disparate discourse communities overlap. This in turn enables political action by diffusing borders, enabling the formation of coalitions by recognizing as well as creating affinities and allies.

The concept of differential consciousness maps on to Black Feminist thought around the construction of homeplace in provocative ways. As bell hooks describes the homeplace, it exists as an island surrounded by hostility. It is a safe refuge that must exist in opposition to the harsh world of its environs and also coexist with it in a state of mutual tolerance and even codependency. The women who construct such places are aware of the balance they must maintain, of creating a place where they can be different and how such self-expression can only occur in an environment that is safe. To be safe a number of things must happen. First, a negotiated tolerance must exist between the homeplace and the harsh world. Second, a selective utilization of resources of the harsh world must occur—the homeplace is not self-sufficient and the people who create it and exist within it are well aware of that. Third, alliances must be forged between people in the homeplace and outside this means negotiating a mutual appreciation not just tolerance. None of this means that either the women constructing the homeplace or the people who live within it sell out; rather it means an opportunistic tolerance and selective appreciation.

Clearly we see Darkside doing this as he embraces science as a tool to enact change in his neighborhood. As science educators, we are particularly interested in how Darkside uses his construction of homeplace to transform the harsh world of his community and of science for himself and his peers. Darkside has clear views about what constitutes science and, in particular science in the community, in both what he says about his work in the action research project and what he does to help sustain the project. As we reflect on Darkside's actions and beliefs alongside the construct of homeplace, we contrast these to science education experiences in schools. This contrast illuminates qualities of homeplace.

Darkside believes that science is important for him to learn. He has signed up for science in school because he believes it will help him get a job. Although Darkside believes that he learns some science in his science class, he does not see learning science in schools as making him part of the scientific community or connecting him to his own community.

Darkside describes school science as a place to learn scientific facts, but that he does not even get to do that because they "dumb-down" and "water-down" the curriculum at his school through the overuse of "worksheets" and "boring assignments" meant to keep inner-city students "in their place." Through his talk and action in the action research project, this is just what Darkside attempts to avoid.

An example involving school science takes place when talking with Darkside and another teenage boy, Goldberg. Darkside is not afraid to inform his peer about the ways in which schooling keeps access to science difficult if not impossible.

Angie: Goldberg, you said, no. Why aren't you going to take science all four years of high school?
Goldberg: Because they might not give it to you.
Angie: It is not offered every year?
Goldberg: No.
Darkside: That's not true! Science is offered every year!
Goldberg: No. It depends.
Angie: On what?
Goldberg: What schedule you get.
Darkside: Or, who's teaching it. But, they just telling you that. You can take science. I know they just telling you that.
Goldberg: But, what if you got teachers that just write on the board and don't explain nothing.
Darkside: Yeah, but you going have to learn on your own then because in college they ain't going to write or explain anything.

Darkside uses his understanding and value of the community and the practices of remembrance and subversion to construct a science that pushes this vision of community forward. Even more interesting, perhaps, is how he implicitly draws on a critique of school science to do so—he places schools and school science in an oppositional position to raise the consciousness of his friends and construct a sense of unity amongst them.

Outside of school, Darkside likes science. In his recent involvement in creating a video documentary on teenage life in the inner city, Darkside has described science as having two domains. The first domain connects to topics to be studied. For example, Darkside has described science as being about the environment and how to keep it from being polluted or about the body and how to keep it strong and healthy. In these descriptions of science he uses talk riddled with scientific terminology such as studying the environment involves "experimenting," "observing," and the "scientific method." The usefulness of science or scientific knowing is what is interesting here and what we see as one of the subtexts of his talk. He wants to learn about the body *and how to keep it strong*. He wants to learn about the environment *and how to keep it from being polluted.* Here is where the second domain of science for Darkside emerges: Science is utilitarian. In Darkside's words, science, and, in particular, science in the community, require the following three qualities:

> It is something that you do in your community that you can be proud of. It is something that you do in your community to be remembered by. And, science is something that will help to beautify and change your community to make it a better place for yourself, your family and your community.

He uses this vision to influence other teens on the action research project. The most poignant example, perhaps, was his idea to create a video by teens for teens about science and the inner-city community. In his words, he wanted to "help other teens see how you can do something for the community." However, the power in Darkside's attempt to make homeplace for his peers comes across in his consistent, but subtle, attempts to built community on a shared history and common anguish.

For example, he uses his "hate" of shelter life to provoke his peers into a passionate stance on their reality and even action. During one particular conversation between two other teenage male peers, Goldberg and Steve, living in the shelter and living elsewhere became the focus of conversation. Although all of the boys agreed that they did not enjoy living in a homeless shelter, one of the boys insisted that eventually he wanted to leave the inner-community all together. Darkside responded by calling attention to their work at the lot, reminding the other boys that it is their job to be proud of their community, and asking them to contribute to it so that it is a place where they want to live and other people want to visit. He told them this was important because it was their community and because he did not want to let the world believe their communities were no good (Fieldnotes, March 2, 1999).

Conclusions

As we reflect on Darkside's experiences, homeplace, and ecofeminism, we are forced to respond to the question, so what does this mean for school science? How might the stories of Darkside, of homeplace and of ecofeminism help us articulate in more vivid and meaningful ways, science for all? Perhaps one of the most important messages embedded within our narrative is how homeplace—a political and radical transformative location—can serve as link between science and social action. In many ways, this position is already supported by our national reform initiatives in science education in the US. For example, the two major goals guiding the "science for all" movement is to promote "science for all Americans" (AAAS 1989, 1993; NRC, 1996; NSTA, 1996), and to translate this vision of Science for All into concrete programs, policies and practices for teachers, curriculum developers, publishers, and students. Here, scientific literacy has been defined by the American Association for Advancement of Science as the

> understandings and habits of mind that enable citizens to grasp the inter-relationships between science, mathematics and technology, to make sense of how the natural and designed worlds work, to think critically and independently, to recognize and weigh alternative explanations of events and design trade-offs, and to deal sensibly with problems that involve evidence, numeric patterns, logical arguments, and uncertainties. (AAAS, 1993, p. xi)

In other words, as Project 2061 suggests, we need to work with students to help them become "users and producers" of science.

Arguing from our position as feminists committed to social change we see embedded within this vision a fundamental orientation to science and education as with/for social action. Clearly the phrase scientific *literacy* implies a degree of knowing and acting upon that knowing. However, what we see as critical to bring out in this discussion is that such knowing and acting upon knowing transcend the academic sphere in order to integrate it with the social sphere. Darkside, through constructing homeplace, shows us that feminism in science education is about constructing places in which the enterprise of science can be rethought and "science" can be placed in a position as a tool for enacting societal change for the better. This in turn means a reconstruction of both "homeplace" of the child's situation and the "harsh world," the science itself. In this sense the "knowing and doing" of science in schools is the active intersection between the lifeworlds of children and the external worlds they come to study. Science education as social action is about pursuing those intentions and intersections and as such is about reconfiguring power relations for we assert that any social action is fundamentally tied to real and ideological shifts in power relations.

Thus, we have argued that central to a construct of science education for social change is giving children the tools and power to do this re-creation. In our story, Darkside and his differentness from our school-scripted versions of scientists, good science students, and science, redefined science, the scientific community, and his role in that process. It is not so much that Darkside radically altered the natural/cultural phenomenon of his local community and the garden lot, but rather, that Darkside altered our conceptions of how these interactions constituted science.

Acknowledgments

Support for this research was partially provided by the National Science Foundation (REC 9733700), and is acknowledged with gratitude.

References

American Association for the Advancement of Science. (1989). *Science for all Americans*. Washington, DC: AAAS Press.

American Association for the Advancement of Science. (1993). *Benchmarks for scientific literacy*. New York: Oxford University Press.

Collins, P. (1990). *Black feminist thought: Knowledge, consciousness and the politics of empowerment*. Boston, MA: Unwin Hyman.

Collins, P. (1998). *Fighting words: Black women and the search for justice (Contradictions of Modernity, v. 7)*. Minneapolis: University of Minnesota Press.

Gaard, G., & Murphy, P. D. (1998). *Ecofeminist literary criticism: Theory, interpretation, pedagogy (The environmental and the human condition)*. Champaign: University of Illinois Press.

Haraway, D. J. (1991). *Simians, cyborgs, and women: The reinvention of nature*. New York: Routledge.

Haraway, D. J. (1992). The promises of monsters: A regenerated politics for inappropriated others. In L. Grossberg, C. Nelsom, & P. Treichler (Eds.), *Cultural studies* (pp. 295–337). New York: Routledge.

hooks, b. (1981). *Ain't I a woman?* Boston, MA: South End Press.

hooks, b. (1984). *Feminist theory: From margin to center*. Boston, MA: South End Press.

hooks, b. (1990). *Yearning: Race, gender, and cultural politics*. Boston, MA: South End Press.

Merchant, C. (1990). *The death of nature: Women, ecology, and the scientific revolution*. San Francisco: Harper.

Minh-ha, T. (1989). *Woman, native, other: Writing postcoloniality and feminism*. Indianapolis, IN: Indianapolis University Press.

Mohanty, C. T. (1991). *Third World women and the politics of feminism*. Indianapolis: Indianapolis University Press.

National Research Council. (1996). *National science education standards*. Washington, DC: National Academy Press.

National Science Teachers Association. (1996). *NSTA Pathways to the science standard*. Author.

Sandoval, C. (1991). U.S. third world feminism: The theory and method of oppositional consciousness in the postmodern world. *Genders, 10*, 1–24.

Sandoval, C. (1995). New sciences: Cyborg feminism and the methodology of the oppressed. In C. H. Gray (Ed.), *The cyborg handbook* (pp. 407–422). New York: Routledge.

Young, I. M. (1990). The ideal of community and the politics of gender. In L. Nicholson (Ed.), *Feminism/postmodernism* (pp. 300–323). New York: Routledge.

9

The Language of Science and the Meaning of Abortion

Nancy Lawrence & Margaret Eisenhart

In the book, *Women's Science: Learning and Succeeding from the Margins* (1998), Margaret Eisenhart and her co-authors, including Nancy Lawrence, examined four sites of non-laboratory science for the meanings of science and of being a scientist that existed there. These sites included a high school genetics classroom, a college engineering internship, an environmental action group, and a conservation agency. The authors showed that although many academic and research scientists are disdainful of the science of such groups, the groups provide important and often overlooked opportunities for learning science. Numerous young people and adults participate in a variety of groups associated with environmental protection, historical preservation, local development, community planning, and political campaigns similar to the ones examined in *Women's Science*. These groups often rely on science and do so in politically charged ways. As Sandra Harding has noted, "Groups with conflicting social agendas have struggled to gain control of the social resources that the sciences—their 'information,' their technologies, and their prestige—can provide" (1991, p. 10). In fact, the success of such groups in contemporary U.S. society often depends on their ability to use science-related information in support of particular political positions. Eisenhart et al.'s evidence suggests that, in consequence, they can be places where people, once discouraged by academic science, become in-

terested in science, and where others, who have previously pursued academic science, can learn more about science.

Eisenhart et al. also found that women were well represented
in these groups and successful at what they did in them. That is,
unlike the familiar profile of women under-represented, unsuccess-
ful, or discouraged in the offices and laboratories of elite research
science, women in these less-elite sites learned and effectively
used the science of the group in numbers proportional to men,
were rewarded accordingly, and enjoyed what they did. In this
article, we examine the science and science learning in two new
groups of (mainly) women who have organized for political effect
(one on each side of the abortion debate). As part of this effort,
they have taken up medical science to advance a cause in which
they deeply believe.

Significantly, the study of the two abortion groups—one a pro-
life group ("PL"), the other a pro-choice group ("PC")—did not
begin as a study of science or science education. Lawrence began
her original study of them as part of her dissertation (under
Eisenhart's direction) on the rhetorical uses of "choice" (as in, for
example, "reproductive choice" and "educational choice") in the
U.S. political campaigns of 1992 (Lawrence, 1994). Like many
before us, we did not expect to find "real science" in these abortion groups. However, when Eisenhart reviewed Lawrence's fieldnotes, she noticed that both groups relied on and discussed a considerable amount of medical science in a context dominated by
women. Both groups were involved in rallying public support for
(opposing sides of) several political campaign issues, lobbying
legislators, holding group meetings to discuss positions and
strategies, and working at abortion clinics (either in support or in
protest of them). Medical science was part of discussions and
strategy in each group. In PL, members relied heavily on medical
science experts to argue that life begins at conception and thus
that abortion is murder. In PC, members relied on (different)
medical science experts to argue that women's lives are jeopard-
ized by illegal abortions or by the inability to obtain an abortion
and thus that abortion should be a woman's right. The PC group
coupled this use of science with a more global one, contending that

abortion was a check on spiraling population growth and consumption of natural resources.

These characteristics of the abortion groups made them similar in several ways to the other groups that were part of the research project that culminated in *Women's Science*; thus they were added to that project. As our data collection and analysis for *Women's Science* proceeded, however, we were discouraged from including the abortion groups. Funders, reviewers, and publishers expressed various concerns about the abortion groups not "adding anything" to the analysis, making the book too long, and so forth. Eventually, Eisenhart gave in and removed the discussion of the abortion groups from the book manuscript (for more about this struggle, see Eisenhart, 2000). In this article, we finally get to tell the story of science in the abortion groups. We continue to believe that science educators and science education researchers are misguided not to be interested in the kinds of science that ordinary people use to make meaning and take action in their lives.

Information about the two groups was collected by Lawrence from group documents, meetings, state legislators (on both sides of the abortion debate), and individual political activists who, for the purposes of this research, were defined as individuals devoting five or more hours per week to promoting or opposing abortion. The study of PL and PC took place over a period of fifteen months, from November 1991 to January 1993. During this time, Lawrence conducted content analyses of each group's documents, attended 15 meetings and conferences held by the groups, observed a half-day state legislative hearing on an abortion bill, and formally interviewed six pro-life activists or state legislators and eight pro-choice activists or legislators. Notably, Lawrence's study took place during an U.S. presidential election year, a time of peak activity for both groups.

The Context of PL and PC

Although readers may (correctly) assume that the science used by these groups was simplified and slanted toward a pre-determined position, the science-related activities of these groups of women should not be casually dismissed for several reasons. For one thing, we found that PL and PC were, in fact, sites for learning science, and the women who participated in these groups took special pride in their increased ability to discuss medical science, which they attributed to participation in the group. Most members identified their increased ability to use medical science to defend the group's position in public as "empowering" and one of the most important consequences of group participation. One PC woman said,

> [My] knowledge of reproductive health care has increased considerably. I've learned things about my body in the past two years [since joining PC] that I didn't know, and I'm 25, and that's really pathetic. . . . I don't understand why I didn't actively seek out this information before, and I don't understand why it wasn't more readily available to me. . . . And I've become much more aware of my menstrual cycle and how it affects me. . . . I'm much more aware of myself physiologically. . . . And that's empowered me as a sexual person. (Lawrence, 1993, pp. 47–48)

A woman in PL, an engineer, was particularly proud of besting a woman from the National Organization of Women (NOW) in a discussion. The PL woman described the incident:

> I did have a big discussion with the woman down at the NOW rally . . . who told me she was one of ten children, but she really didn't have any empathy for this . . . bunch of cells in the womb. Now this is some kind of college-educated woman who's been deceived somehow or they never got into the study of the fetus. I mean, you go to any library and get a book . . . and they'll show you, stage by stage, the finger development and everything, the ear development. (Lawrence, 1993, p. 20)

Men, too, felt they had been affected by the science they learned in the group. One, an atmospheric scientist in PL, was asked whether he was familiar with fetal development before his PL involvement. He said:

> No, I learned it [about fetal development] as I became active. The more I learned, the more active I became. You know, when you get in there and look, and you say, "Hey, I once was that little person." Now, something didn't change along the way that made the individual from a fetus to you and me. I mean, that just doesn't happen. You and I and the head of

Planned Parenthood were once one of those in the womb. (Lawrence, 1994, p. 133)

In these examples, well-educated adults were discussing the power of the science presented to them in PL and PC. Regardless of the "accuracy" of this science, group members were motivated to learn the group's science, felt good about what they had learned, and expressed an interest in learning more.

Further, they intended to use this knowledge to increase the credibility of their own positions in public and to influence the ideas and actions of others. In other words, PL and PC made scientific knowledge worth having, especially for its use in public discussion and debate.

A second reason for those interested in science and gender to give serious consideration to PL and PC is the fact that the two groups are putting scientific knowledge to use in the political service of women-focused, reproductive issues.[1] Historically, this is a rare convergence, because women have been systematically excluded from both politics and science. In 1976, Lorenne Clark, writing about women's exclusion from politics, said:

> At its very heart politics assumes that reproduction is demeaning, that fundamental meaning and importance for man . . . cannot derive from the genesis and nurturance of children. The idea of politics, and the theory to support it, was born from a sense of the futility of the reproductive function. It was designed exclusively by men, exclusively for men, as an escape from the world of the household, the realm of the "merely" biological. (Clark, 1976, p. 51)

In a similar vein, Jane Roland Martin (1994; orig. 1981) has argued that since Plato, the leading images of an "educated person" and a "citizen" have been prototypically male, making it a contradiction for women to be either. With regard to science, feminist researchers have discovered numerous ways in which women have historically been excluded and ignored (see, for example, Harding, 1991). In 1989, Martin summed up the long-standing biases against women in science as follows:

---

1. Certainly, reproductive issues involve men; however, historically and up to the present time, reproduction has been culturally associated with women.

[The] charges of bias [in science fields] range, in turn, from claims about the composition of the profession and treatment of females in it to claims about the failure of the profession to investigate topics of particular concern to girls and women; from the ways in which scientific research victimizes women, to those in which scientists have projected their culture's gender stereotypes onto the natural world; from characterizations of the language used in our scientific theories as masculine to claims that our very thinking about science and nature is genderized. (Martin, 1989, p. 242)

The fact that well-educated women are working actively in PL and PC to affect the ways reproduction is treated in public life, and doing so by recourse in part to science, represents a significant change in women's political activism and the landscape of movement politics. And, the change is not likely to end with this single occurrence. James Davison Hunter (1994), in his analysis of America's "culture war," argues that a newly emerging "politics of the body" (see also Foucault, 1990; Ross, 1991)—with abortion as the first well-developed U.S. example—will become *the* dominant focus of U.S. public debate in the foreseeable future. If Hunter's analysis is correct, then the use of science by women in PL and PC should not be considered unusual or bizarre, but a harbinger of things to come in the political struggles of the next decades.

Nonetheless, PL and PC presented some obstacles to women. A major one was that the groups' political agenda constrained expression of the complexities and ambiguities that women felt about abortion and obscured their common ground (Lawrence, 1994; see also Maloy & Patterson, 1992). The personal expressions of PL and PC women were, in fact, quite similar. For example, both PL and PC women talked at length (in personal interviews) about poor mothers' difficulties in providing for their children. Both PL and PC women worried that, as a society, the US was not offering mothers the support they needed to raise their children well. PL and PC women also shared a dislike of the need for abortion and a concern that fathers' perspectives were not given adequate consideration. Yet, in the activities of the PL and PC groups, these issues and the complexities they add to the abortion debate were rarely included. All the women activists we knew said that "political realities" (perhaps especially in an election year) precluded the groups from dealing with these issues:

Everyone seemed to accept that, to be seen and heard in the current political climate, the groups must stake their positions on other, opposing concerns.

Maloy and Patterson (1992) suggest that this finding is not unique to the groups or specific time frame we studied. In their own study of pro-choice and pro-life groups, they found:

> The heat of the conflict has, for all useful purposes, reduced a great many probing and subtle thoughts on the subject of abortion to just two antithetical positions. While thoughtful scholars, medical professionals, and conscientious citizens labor to explore and understand the issues, advocates of the inadequately labeled pro-choice and pro-life... positions have been able to capture the spotlight—whose glare has bleached any shade of gray. These two factions have come to represent abortion in deceptive black-or-white terms. Their arguments are directed toward preserving or overthrowing existing law, not exploring in depth the truths and quandaries that surround the issue. As a consequence, little substantive dialogue has taken place in recent decades in the media or in legislatures. Rather, the rush to win votes, rights, or restrictions has clouded the controversy with hostility and even violence. (Maloy & Patterson, 1992, pp. 1–2)

Appeals to science contributed both to the progressive potential and the limitations in the pro-life and pro-choice groups Lawrence studied.

Inside PL and PC

Analysis of the data collected from both groups reveals that appeals to science prove useful both to those who would prohibit abortion and those who seek to keep abortion legal. Frequently portrayed as an impartial and dispassionate discipline, science is used by both groups to persuade and to strike emotional chords.

The Pro-Life Group

The language of science is at the center of two books, *Pro Life Answers to Pro Choice Arguments* by Randy Alcorn and *Handbook on Abortion* by Dr. and Mrs. J. C. Wilke. The *Handbook on Abortion* is hailed by the pro-life movement for debunking abortion myths and arguments of abortion rights groups and referred to frequently in the PL group we studied. The first book is a 1992 publication produced by Multnomah Press, a ministry of Multnomah School of the Bible. The second, and more widely known book, is one of the earliest books produced by the pro-life movement. First published in 1971, it has seen numerous printings and several revisions.

Identifying the "science strategy" that must be adopted by pro-life groups if their agenda is to be advanced and adopted by a religiously diverse American public, Dr. Wilke writes in the foreword of *Handbook on Abortion*:

> Our emphasis, we are convinced, must be on the scientific, medical and social aspects of this issue if we hope to present the facts in a way that can influence our pluralistic society. Theological considerations are critical to each person individually but cannot be imposed upon other non-believers in the culture. (p. v)

Thus, it would appear that, from the outset, Wilke's goal in writing the book is to use science to spread PL's message beyond those who accept it for religious reasons.

Wilke then takes up the question "Is this human life?" He writes:

> Back to our basic question. Is this unborn being, growing within the mother, a human life? Make this judgment with the utmost care, scientific precision, and honesty. Upon it may hinge much of the basic freedom of man in the years to come. Judge it to be a mass of cells, a piece of meat? Then vote for abortion-on-demand. Judge it to be a human life? Then join us in fighting for his and her right to live, with all the energy and resources at your command. (p. 6)

A medical doctor himself, Wilke relies on other scientific and medical "authorities" to corroborate his statements on the beginning of human life. In a section titled "What is the opinion of natural scientists?" Wilke notes that a "distinguished scientific meeting" in 1967 brought together "authorities from around the world in the fields of medicine, law, ethics, and the social sciences at the First International Conference on Abortion to consider the question, 'When does human life begin?'" (p. 7–8). Wilke includes the group's "almost unanimous conclusion (19 to 1)":

> The majority of our group could find no point in time between the union of sperm and egg, or at least the blastocyst stage, and the birth of the infant at which point we could say that this was not a human life. . . . The changes occurring between implantation, a six-week old embryo, a six months fetus, a one-week old child or a mature adult are merely stages of development and maturation. (p. 8)

Randy Alcorn, author of *Pro Life Answers to Pro Choice Arguments*, is not a medical doctor or a scientist. However, he preempts criticism of his lack of formal science training by relying on doctors and scientists as sources. Included in his 16 acknowledgments are two MDs, three RNs, two Ph.D.s, a former abortion clinic owner, and a former abortion clinic counselor.

Like Wilke, Alcorn anchors his arguments in the words of medical doctors and scientists. Heeding Wilke's call to emphasize the scientific and medical aspects of abortion to persuade Americans that abortion is wrong, Alcorn rebuts the statement. He argues, "It is uncertain when human life begins; that's a religious question that cannot be answered by science" by reference to scientific consensus. Thus, "Medical textbooks and scientific reference works consistently agree that human life begins at conception" (p. 39).

Alcorn's scientific assertions are supported by quotes from doctors and scientists in the fields of biology, embryology, pediatrics and obstetrics, pathology, and genetics. Alcorn reminds his readers that the doctors' remarks are secular and scientific and not religious. Noting that his sources arrived at their conclusions from scientific facts, Alcorn writes:

> These sources confidently affirm, with no hint of uncertainty, that life begins at conception. They state not a theory or hypothesis and certainly not a religious belief—every one is a secular source. Their conclusion is squarely based on the scientific and medical facts. (p. 40)

Both Wilke and Alcorn work scrupulously to emphasize that their arguments against abortion proceed from scientific fact and not religious conviction. According to Alcorn, "The pro-life position is based on documented facts and empirical evidence" (p. 200). The message (supported by science) is that abortion is wrong because abortion "stops a beating heart." By invoking science, these pro-life advocates hope to attract and persuade a wider audience, reaching beyond pro-life religious groups.

Another powerful and more emotional use of science is represented in the photographs contained in Wilke's and Alcorn's books. The photographs are of embryos, each with its own caption, including: "preborn baby," "unborn child," "feet of miscarried child," "preborn baby killed by saline abortion," and "hand taken from the discarded remains of an abortion." Wilke's color photos are more graphic than Alcorn's and include dismembered fetuses with these captions: "scraping," "suction at 10 weeks," and "salt poisoning at 19 weeks" (pp. 26, 27). (The pictures are credited to a Canadian pathologist and a Maryland doctor.)

Drawing an analogy between fetal pictures, the anti-smoking campaign, and animal rights, Alcorn states:

> Banning such pictures from the abortion debate is like banning x-rays of smoke-damaged lungs from the smoking debate, or saying we cannot show pictures of harpooned whales when discussing animal rights. (p. 187)

Alcorn concludes:

> If the fetus is simply a lump of tissue, then fine—let the public see the pictures of the lump of tissue. Let them be treated like adults and allowed to choose for themselves what they believe. If this is not a baby, what could be the harm in looking at the pictures? (p. 187)

Examination of pro-life literature reveals that the use of science, through words and pictures and corroborated by medical doctors, is central to the group's strategy of persuasion. Their scientific argument is intended to reach an American public not persuaded by religious and moral claims that abortion is wrong. Their central message is backed by "actual" pictures and carefully chosen labels intended to at least startle if not enrage viewers.

The language of science was also at the center of a controversial abortion bill—favored by PL and opposed by PC—heard during the 1992 Colorado legislative session. The bill, Senate Bill 169, was known and referred to as the "Woman's Right to Know

Act." Had it passed, SB 169 would have required "relevant information be provided to a woman prior to the woman's choice whether to have an abortion" (Senate Bill 92-169, p. 1). Additionally, the bill included a 72-hour waiting period to "assure that each woman's decision regarding the abortion procedure be based on an opportunity to reflect upon the information received" (SB 92-169, p. 3). The "information" included:

> the nature and description of the various abortion procedures and the physical and physiological risks associated with such procedures, including any danger with respect to future pregnancies; the anatomical and physiological characteristics of the unborn offspring covering two week gestational intervals from fertilization to full term, including photographic replication for each two week gestational interval. (SB 92-169, p. 9)

The "Woman's Right to Know Act" was heard on February 7, 1992. The hearing opened to a packed room of Colorado citizens, journalists, activists, and school children from area civics classes. The 14 witnesses (seven per side) shared similar titles and degrees. They were medical doctors, nurses, educators, professors, attorneys, activists, and private citizens. Their support or opposition to SB 169 marked their differences.

Witnesses supporting SB 169 focused their testimonies on "informed choice," emphasizing that a woman considering abortion should be very "knowledgeable" about fetal development and various abortion procedures before undergoing an abortion. Additionally, supporters claimed that SB 169 would "empower" a woman by providing her with scientific facts and knowledge. Identifying the reason behind his introduction of the bill, the bill's sponsor stated:

> The purpose of this bill is to empower women. It calls for an information packet, [designed] solely by the state health department. [Containing] factual and scientific information, [describing] various abortion procedures.

The first witness to speak in support of SB 169 was a male attorney from a Washington, D.C., law firm. His defense of SB 169 proved to be a theme expressed by subsequent witnesses in favor of the bill: the reasonableness of the bill as a means of providing scientific information to pregnant women. The attorney testified:

> Let's pretend this is a bill requiring information about breast cancer. California has such a bill. It outlines risks of various treatments; advantages and disadvantages of treatments, and discusses alternatives.... In Massachusetts, patients must receive information about the number of spinals, epidurals, C-sections.... In short, outside of the abortion context, there is nothing unique about informed consent. It is the medical issues that should make your decision today not the political issues.

In general, the strategy used by proponents of SB 169 cast the "Woman's Right to Know Act" as providing medical information essential in making an "informed choice" about abortion. Individuals considering abortions should make the decision to terminate their pregnancies informed by "knowledge," "facts," and "truth."

Science was also prominent in several workshops presented at the second annual "Pro-Life Conference," sponsored by a county Right to Life Committee and attended by PL members. Of the four sessions attended by Lawrence (Voter ID, Apologetics, Feminism, and The Caucus System), the Apologetics workshop invoked the language of science most heavily. Facilitated by a member of Citizens for Responsible Government, a statewide pro-life organization, the leader began by offering the audience a definition of apologetics: "Apologetics is a formal defense of a belief or cause." Thus the goal of this workshop would be to articulate a well-conceived position.

The speaker noted that "abortion is framed" in several ways: as a biological issue, as a human rights issue, and as a legal issue. He said it is also framed as "a moral and religious issue but this is difficult to debate." Admitting this last point, the speaker urged the audience to think of abortion in biological terms when presenting the pro-life message. Cautioned the speaker, "You don't want to carry your coffee table bible into the debate with you."

The discourse within the PL group focused primarily on how best to reach, i.e., educate, diverse groups with the PL message. Said one male activist: "It's a non-political group; it can't endorse candidates ... it's an educational institution ... and is educational in its primary function." When asked how his group "educates," he used the same scientific language favored by other PL activists: "Well, when we have the opportunity, we describe the nature of the unborn child, development in the womb, the scientific facts of reproduction." Further, at their first meeting following the November 1992 elections, members of the group expressed their disappointment that pro-life candidates lost so overwhelmingly in local, state, and national races. Said one member, "Education will become the most important thing now with the outcome of the elections.... We have to focus more on education."

Nodding in agreement was the only male present, Don, an atmospheric scientist at the local university. His remarks that night echoed earlier comments he made during an interview, "Younger kids, junior high kids, they're still learning about life. You show them it's a baby growing. They're not fooled. They know it's life." There was agreement that students must be educated about "when life begins," and it was suggested that the organization put advertisements in the papers of a local high school, "Can we put this in the ... high school paper? ... Should we try it in January? January is coming up again."

There were slight groans when January was mentioned, for on January 22, 1993, the anniversary of *Roe v. Wade*, abortion rights activists would celebrate 20 years of legal abortion. A member suggested that the organization run an advertisement in the local daily newspaper.

Discussion turned to what kind of advertisement to put in the paper. There was unanimous agreement that the ad should not be violent, because a picture of an aborted fetus might go too far in angering readers. The president circulated two leaflets printed on slick paper with delicate photographs of a "19 week unborn baby." The caption of one read, "The incredible photograph above by Dr. Rainer Jonas shows what a healthy, active intrauterine child looks like at 19 weeks. Like the bud of a flower, beautiful. But, unfortunately, still a candidate for elective abortion." Words in large, bold print above the photo, read, "When they tell you that abortion is a matter just between a woman and her doctor, they're forgetting someone. There are alternatives to abortion. There have to be." The second photo contained the question, "Is this a choice? Or a child? There are alternatives to abortion. There have to be." A third leaflet depicted a fetus with its thumb in its mouth but no indication of age. These words appeared beside the photograph: "If he is not alive, why is he growing? If he is not a human being, what kind of being is he? If he is not a child, why is he sucking his thumb? If he is a living, human child, why is it legal to kill him?"

It was this last photo that was eventually selected to run in the local newspaper. While there was no indication of age, the photo of the fetus, enclosed in its amniotic sac, looked distinctly human. The photo captured those parts of the fetus that most re-semble a newborn baby, including a delicate ear, curled fingers, two fragile shoulders, a thin chest, a closed eye, and a tiny head covered with a complex weave of veins and vessels. Revealingly, the photo did not contain the entire fetus. Not visible was any-thing below mid-chest, including the umbilical cord, the most fa-miliar reminder that a fetus is still in utero.

Findings from talk during PL meetings suggest that most of it is devoted to various means of broadly communicating PL's main message—a message defined in large part by the PL literature in which so-called "scientific facts" about human life, conception, and fetal development have a prominent role. At least for the group we studied, "educating the public" about PL's "scientific" position was the paramount concern. To accomplish this, the PL literature, as well as group members, repeatedly emphasizes the words and statements of "medical experts" and "research scien-tists" and uses them to try to convince the public that PL is on the side of "science."

The Pro-Choice Group

The pro-choice (PC) group is "a political action group dedicated to all issues affecting reproductive health" (*Update*, PC newsletter, 1993). Formed in 1991, PC is an arm of a local, state, and na-tional pro-choice organization, an organization committed to im-proving "the quality of life by enabling all people voluntarily to exercise individual choice in their own fertility and reproductive health" (1992 brochure). While PC and larger pro-choice organizations unwaveringly defend reproductive choice, abortion is but one component in their broad mission championing and providing reproductive health care.

The literature used by the group included *Happenings*, a political action newsletter published eight times per year by the state organization, and *Update*, an every other month publication of the state organization. Also available were fund-raising letters, pamphlets and brochures produced by the state organization, and literature published by the Religious Coalition for Abortion Rights (RCAR), an organization of 35 national Protestant, Jewish, and other religious groups.

An example of the role of science language in pro-choice literature appears in a 1992 *Happenings* newsletter. The issue included an insert describing the abortion pill RU486, and called on readers to "act" to make the French-manufactured abortion pill available in the United States:

> Pressure to make RU486 available to women in this country and overseas must continue and accelerate. It is critical to press forward on the scientific and technological front while at the same time fighting on the political and legal front. Anti-abortion extremists must be exposed for what they are—anti-science and anti-birth control. It is shocking and intolerable that a drug hailed by doctors around the world as the most significant discovery in reproductive medicine since the oral contraceptive is barely being researched in the United States.

Calling for an all-out effort from pro-choice activists to make RU486 available, the statement continues:

> Such an effort must grow and reflect the mainstream of the American public in order to assure that an extremist minority does not succeed in blocking scientific progress and inhibiting a medical advance that is optimistic, far reaching and promises a more safe and healthy future. The message is simple: to be pro-RU486 is to be pro-science and pro-health. Conversely, to be anti-RU486 is anti-science and anti-birth control.

This example reveals how PC attempts to align itself with science. Calling those who oppose RU486 "anti-science," against a "most significant discovery," and opposed to a "medical advance," this literature attempts to depict PC on the side of rational, scientific

progress. In this literature, abortion becomes a procedure characterized as a medical and scientific "advance."

PC literature also offers numerous statistics, attributed to research studies, to support its positions. The following appeared in
a March 1992 *Happenings* under the title, "AMA Report Cites Reduction in Abortion Related Deaths Since *Roe*":

> A new study has found a sharp reduction in abortion-related deaths
> since the Supreme Court's Roe v. Wade decision legalized abortion in
> 1973. The study also showed that the risks of dying from pregnancy and
> childbirth are nearly 12 times as great as the risks of dying from an abor-
> tion. [No reference information provided.]

Statistics are used to suggest scientific and medical justification
for legal abortion. As evidence that illegal abortions are medically
dangerous to women, statistics are cited to persuade the reader
that legal abortion protects women. Indirectly responding to the
"abortion kills fetuses" charge leveled by the pro-life group, the
pro-choice group claims that "Laws against abortion kill women."
The use of statistics, coupled with emotional words like "dangerous," "secret," "illegal," "mutilated," and "criminals" paint an
ugly picture of life before *Roe*:

> To prohibit abortions does not stop them. When women feel it is abso-
> lutely necessary, they will choose to have abortions, even in secret,
> without medical care, in dangerous circumstances. In the two decades be-
> fore abortion was legal in the U.S., it's been estimated that nearly a mil-
> lion women per year sought out illegal abortions. Thousands died. Tens
> of thousands were mutilated. All were forced to behave as if they were
> criminals. (Nine Reasons Why Abortions Are Legal, 1990, p. 2)

The statistics tell a grim story and ask the question: Does the
American public want legal, and thus medically safe, abortions or
will abortions be prohibited, making the procedure, and potentially childbirth, medically dangerous?

While the pro-choice organization rarely responds with
graphic pictures, they do address the use of such pictures by the
pro-life movement. During the 1992 U.S. Senate campaign in Colorado, a pro-life candidate aired television ads showing "aborted
fetuses." In an October 1992 issue of *Update*, the TV ads were
called "dishonest" and "insensitive." Disputing the credibility of

the ads, the pro-choice group responded with statistics and their own fetal developmental evidence:

> The ads show dead fetuses which [U.S. Senate candidate] claims are the result of everyday abortions. It is evident from their size and stage of development that these fetuses are nearly full-term and may be the result of third trimester stillbirths. They clearly do not represent first trimester abortions, which comprise 91 percent of all abortions performed in the United States.

In this example, the pro-choice group challenges the scientific facts of the "aborted fetuses" ad campaign and replaces the pro-life "facts" with their own pro-choice "facts."

One other use of science by the pro-choice group is evidenced in a document that received legislative and public attention during testimony against Colorado's SB 169, the "Woman's Right to Know" measure. Testifying against SB 169, a witness from the state pro-choice organization was asked if her group provided information to women seeking abortions. The witness replied in part, "An abortion facts book is given to all women."

The *Facts Book for Abortion Patients*, more accurately described as a pamphlet, is a procedural document describing the steps before, during, and following an abortion. The "abortion procedure" is described as a five-step process, "which takes about 10 minutes and is usually done with a local anesthetic" (p. 3).

The *Facts Book* is written for women who have decided to have an abortion. It conveys procedural and medical information about abortion. The information contained in the document is presented without emotion and using impersonal language. For example, the following sentence appears in the preface: "In this *Facts Book* you will find the necessary information concerning the procedure and your health care following it." The sentence suggests that the material presented in the booklet is accurate and complete. The abortion procedure is described as follows: "The tube is attached to an aspirator or suction machine, which empties the uterus by gentle suction. . . . When the uterus is emptied, a spoon shaped instrument is inserted to carefully scrape the walls of the uterus to check for any remaining tissue" (p. 6). With such language, *The Facts Book* attempts to achieve an air of unbiased, technical truth.

Science was salient in PC meetings but always integrated with discussions of political education, legal decisions, and strategies for achieving election victories. During the meetings, the language of science, in the form of women's medical and emotional safety, contraceptive education and technology, sex education, and population control, was used to emphasize the importance of increased political activism.

The PC group met monthly. One month would feature a guest speaker on a topic related to reproductive rights, birth control, and abortion. Alternate monthly meetings were without speakers, and discussions centered on political education and strategies and legal updates and responses. One of the guest speakers in 1992 was a candidate for the state House.

Speaking to PC less than a month before the general election, the candidate made her case for abortion rights by reference to statistics on population growth.

> In 1850 it took the entire history to get to one million. By 1930 the population had doubled to 2 billion. We're 5.5 billion in 1992. We won't level out until the middle of the next century. It will level out. There's a population momentum. More and more women are having fewer and fewer children. But there are many women younger than those of childbearing age. Thirty-five percent of the world's population is under 15 years of age. For 12 years, the Reagan-Bush administrations have been concerned about abortion as a political issue. Hopefully, if we have a new administration, we'll take more responsibility. Women are crying for contraceptives. . . . Mexico in the year 2000: 26.3 million. Sao Paolo, Brazil will have 24 million. It's partly a political problem. We can't talk about population in Africa without talking about AIDS. Access to family planning in Latin America, the church and government are starting to look the other way. The total children per women in Latin America has dropped, 5.4 to 4.2. In Africa . . . women are having fewer children. . . . It used to be you had to have seven or eight children to have four survive. A strong ethic. Cultural things as well. As long as your child can say your name, you will live eternally. Twenty-five percent use of contraceptives in Kenya. Other countries it's around 10 percent. [There will be] 180 million children under five years of age in Africa by the year 2025 if the current growth rate continues.

Using statistics on population growth, she emphasized the global consequences—economic hardship, environmental stress, unhealthy families, unchecked urbanization, and the continued spread of AIDS—of limiting abortion rights.

As a rule, in the PC group and their literature, scientific information is used to stress that abortion is medically, environmentally, and personally sound. Statistics from research studies are used to support this position. Impersonal, technical, and procedural language is used to define abortion and describe the process. As was also the case in PL, the PC literature and group meetings Lawrence observed included relatively little discussion of the women's actual experiences of abortion or the beliefs and concerns they actually held about this issue. In general, the role of science in PC meetings was to encourage political activism. Science, both personal and global, was used to underscore the urgency of immediate attention and individual activism.

Discussion

In summary, PL and PC were places where women engaged in science-related activities with more interest and enthusiasm than usually attributed to women in school science. They were also places where women became more politically active on behalf of a cause they cared about. Further, the motivation to learn more about science seemed to derive from a desire to appear credible in public discussions or debates about a politically controversial issue.

In the two groups, gender-focused talk was, of course, salient. The political and personal agenda of people in both groups concerned the prerogatives and responsibilities of women. However, most of this talk was formulaic, that is, phrased and couched in the oppositional, political rhetoric of the pro-life and pro-choice movements. Most of the talk also was used in the service of political activism—in the case of PL, to extend the reach of its message; in the case of PC, to hammer home the dire consequences of a ban on abortion.

In this process, both groups drew on the language of science. In so doing, they accomplished two things. First, they turned woman-centered and locally shared concerns into two opposing positions, each supported by different so-called "scientific facts." Second, by repeated reference to the "facts" supporting their positions, they suggested that their causes were credibly supported, unbiased, and "serious." This process creates and maintains an artificial dichotomy that precludes or preempts thoughtful deliberations or solutions about an important contemporary issue, and it shortchanges participants' interest in learning more science.

Sandra Harding (1991) has argued that "adequate science" must entail accurate descriptions and explanations of the world we (all) live in and the choices we really have. She hoped that by "thinking" via women's perspectives and lives, the practices and theories of science could be expanded, improved, and made more adequate. Unfortunately, this is not what the pro-life and pro-choice groups we studied were enabling women to do with science. Morowitz and Trefil (1992) showed that it is possible to engage in substantive discussions of science that inform the abortion controversy. Surely it is also possible to engage in substantive discussions of women's experiences and concerns so as to inform science. It is unfortunate that despite the importance of the pro-life and pro-choice women's political activism and despite the significance of their appropriation of science, the groups were not also a context for the development of more adequate science for and about women.

What remains important about PL and PC is the motivating context that both groups provide for people, especially women, to learn more science. Women initially joined the two groups because they believed strongly in an issue and wished to be identified with political action related to it. As the women became more involved in the groups, they were motivated to learn more about the science behind the issue that concerned them. This science was not cordoned off as a separate pursuit but integrated with the group's political processes in order to recruit supporters, convince voters, defend positions, and advance agendas. Although the women did not learn as much science, or as sophisticated science, as we might like, they became highly motivated to learn and use the science of the group. At the same time, they were empowered to engage in more social and political activism.

Our case studies of the two abortion groups underscore that the laboratory or research scientist is only one kind of person who is interested in or needs a command of science in contemporary U.S. society. Laboratory scientists use scientific knowledge primarily to further basic knowledge, and they may do so with little attention to social or political implications. In contrast, ordinary citizens face issues of how to obtain accurate scientific information, how to make informed decisions about socially and politically relevant issues, and how to engage in debate about social and political matters in constructive and responsible ways. The quality of the science learned in PL and PC is not what science educators should want, but the motivational context for learning science that groups like PL and PC offer is impressive. For these reasons, the case of PL and PC is instructive for educators who wish to find ways of creating similar contexts in schools, where the content of what is learned could be more fully and satisfactorily explored.

References

Alcorn, R. (1992). *Pro life answers to pro choice arguments*. Portland, OR: Multnomah Press.

Clark, L. (1976). The rights of women: The theory and practice of the ideology of male supremacy. In W. R. Shea & J. King-Farlow (Eds.), *Contemporary issues in political philosophy* (pp. 49–65). New York: Science History Publications.

Eisenhart, M. (2000). Boundaries and selves in the making of 'science.' *Research in Science Education, 30,* 43–56.

Eisenhart, M., & Finkel, E. with L. Behm, N. Lawrence, & K. Tonso (1998). *Women's science: Learning and succeeding from the margins*. Chicago: University of Chicago Press.

Foucault, M. (1990). *The history of sexuality. Volume* 1. New York: Vintage Press.

Harding, S. (1991). *Whose science? Whose knowledge? Thinking from women's lives*. Ithaca, NY: Cornell University Press.

Hunter, J. D. (1994). *Before the shooting begins: Searching for democracy in America's culture wars.* Cambridge, England: Cambridge University Press.

Lawrence, N. R. (1993, April). *The language of science and the meaning of abortion.* Paper presented at the Annual Meeting of the American Educational Research Association, Atlanta, GA.

Lawrence, N. R. (1994). *The choice of language and the language of choice: Public/private discourse about abortion and education in the early 1990s.* Unpublished doctoral Dissertation, University of Colorado, Boulder.

Maloy, K., & Patterson, M. (1992). *Birth or abortion? Private struggles in a political world.* New York: Plenum Press.

Martin, J. R. (1989). What should science education do about the gender bias in science? In D. Herget (Ed.), *The history and philosophy of science in science teaching* (pp. 242–255). Tallahassee: Florida State University Press.

Martin, J. R. (1994; orig. 1981). Sophie and Emile: A case study of sex bias in the history of educational thought. In J. R. Martin (Ed.), *Changing the educational landscape: Philosophy, women, and curriculum* (pp. 53–69). New York: Routledge.

Morowitz, H., & Trefil, J. (1992). *The facts of life: Science and the abortion controversy.* New York: Oxford University Press.

Ross, A. (1991). *Strange weather: Culture, science, and technology in the age of limits.* London: Verso.

Wilke, J., & Wilke, J. C. (1971). *Handbook on abortion.* Cincinnati, OH: Hiltz Publishing.

CLUSTER III

Science Education as (Subversive) Epistemological Praxis

Conversation 3

Michael: In these last three chapters, all authors question traditional science educators' notions of what it means to know science and what the nature of this science is. I still remember the moment Angie Barton was asked during a conference on research in science teaching where the science was in her work with Darkside and his colleagues at the shelter for homeless people and families. These science educators would therefore not agree with the positions laid out in the last three chapters.

Jacques: I remember that instance with Angie Barton, but then, who has the epistemological authority to tell the "truth about the truth"? We will likely come to the conclusion that matters of epistemology cannot be dissociated from matters of power.

Michael: I agree but I think that this connection is most of the time overlooked due to the fact that for a lot of people epistemological practices are conceived as esoteric enterprises that professional philosophers are the only ones to master. They would discuss such matters as principles of demarcation or the status of experiments in producing scientific knowledge without ever leaving the comfort of their university offices. But times have changed ever since anthropologists and sociologists have started following scientists around in their daily activities, showing that scientific research is much less heroic than philosophers of science previously claimed. The social character of scientific knowledge as any other form of knowledge became quite apparent.

Jacques: But I must add that this does not lead us to a position where "anything goes." For instance, it is useful or viable to agree that the concept of electromagnetic waves discussed by the students in your chapter allows us to realize a certain number of projects such as communicating through telephone services. But in principle all knowledge can be interrogated or deconstructed even though we establish closures and take for granted a minimal base of discussion.

Michael: This is where epistemology gets to be very interesting, epistemology thought in terms of practice. In fact who decides what is to be taken as granted? Since there is no ultimate instance we can refer to for making a decision, we again see that the solution to this problem involves power relationships. In a democratic society we have to make sure that the greatest number of people take part in the process.

Jacques: Opening up the discussion does not by necessity lead to an agreement, and therefore ethics comes into play. For instance, when Aikenhead earlier says that knowledge of First Nations people has to be considered on equal grounds, he makes an ethical statement because he then considers the knowledge of the Other equal in value to Western scientific knowledge. What is proper knowledge in Aikenhead's curriculum is the result of a negotiation between different stakeholders.

Michael: In a certain way we can see how epistemology as a practice can become a resource for social action. This is the reason why I not only advocate but also actively include epistemology into my own science teaching.

Jacques: This is quite possible as the next chapters illustrate in many ways, and I hope that, by the same token, epistemology will be demystified in the eyes of our readers since as a practice it can be enacted by very young children.

Michael: As important as this may be, our ultimate goal ought to be a kind of generalization of the epistemological posture whereby it would become "normal" to constantly interrogate some aspect of our knowledge, be it scientific or everyday knowledge. Such interrogation would contribute more to the health of democracy than any course on politics.

10

Science Education as Exercise in Disciplining Versus a Practice of/for Social Empowerment

Marie Larochelle

Now more than thirty years ago, Foucault (1966) began *Les mots et les choses* with a classification of animals which might well induce a kind of epistemological vertigo. This classification, borrowed from Borges, who apparently had borrowed it from a certain Chinese encyclopedia, suggested that animals could be divided as follows:

> (a) those that belong to the Emperor, (b) embalmed ones, (c) those that are trained, (d) suckling pigs, (e) mermaids, (f) fabulous ones, (g) stray dogs, (h) those that are included in this classification, (i) those that tremble as if they were mad, (j) innumerable ones, (k) those that are drawn with a very fine camel's hair brush, (l) others, (m) those that have just broken the flower vase, (n) those that resemble flies from a distance. (p. 7)

The classification is all the more outlandish in that the seemingly familiar (alphabetical) format used to present elements also short-circuits our intellectual codes and habits. How are we to establish an order of things or an overarching category on the basis of categories that juxtapose mythical beasts, real animals, embalmed animals, or animals that have just broken the flower vase? Furthermore, how can we establish such an order on the basis of categories that legitimate (to our way of thinking) incongruous and paradoxical items, such as category "those that are included in this classification" (h)?

Obviously, we are not confronted every day with such unthinkable classifications. However, in a less spectacular and (apparently) less vertiginous fashion, everyday science classes propose distinctions, formulae, and discipline-bound classifications
that may be equally outlandish when viewed in terms of the experience-based knowledge of students and the issues of concern to
them. Indeed, such distinctions and classifications may give students the impression of being confronted with the unthinkable, as
evidenced by the comments of the following secondary school
student:

> They say p equals v times this, you know, and then it's really like it
> seems like almost imaginary.... *I can't picture it, it's not tangible....* When
> we were doing the wave or electricity, the current goes through there. *I
> can't really imagine current going through a wire and stuff like this ... it
> doesn't seem real.* (Roth & Roychoudhury, 1993, p. 34, my emphasis)

The question arises: What is made of this "unthinkable" something? How do the protagonists of the classroom situation come
to grips with it? Is there room for a process of enculturating the
"unthinkable" or for establishing a space in which to bring off negotiations between the order of things being taught and the order
of things "being experienced?" Is there room for hybridizing the
world of scientists with that of students? Or is it simply a question of disqualifying the world of students, even if students have
good reason to think the way they do?

In reference to studies of science teaching in everyday contexts
(Lemke, 1993), and the popular mediation of science (Lévy-
Leblond, 1984; Millar & Wynne, 1988), it would appear that such
issues are often not thought through. This does not mean, however, that these issues have no impact on ways of appropriating
scientific knowledge or on the status with which this knowledge
and those who produce it are imbued. That, at least, is the assumption that underlies the reflections to which I devote myself
here. In particular, I would like to examine the socioepistemological problems that are entailed by this unreflexive process of socialization into a particular order of things, such as scientific
knowledge and classifications.

As an illustration of what I mean, I will examine two cases of scientific classification that are alike for having generated a "feeling of strangeness" or an "epistemological clash." One of these occurred in a primary school setting, and the other among members of the scientific community. I will attempt to show that, depending on the way this clash is appraised, its "epistemological potential" is not the same. Thus, in the case of the scientific community, an epistemological clash throws a wrench into an otherwise seamless thinking process (to paraphrase Lévy-Leblond, 1984). These protagonists can no longer make do with asserting the "truth" of their respective classifications but must now engage in a series of trials, debates, and negotiations over their resistances. But it is an entirely different situation in the classroom. There, the stage has already been set, and the actors' lines have already been scripted. The clash has been ushered back into the wings before the curtain has even gone up, so to speak. In short, students are deprived of the possibility of learning to engage in a reflexive process of testing and negotiating their particular practice of classifying versus that of scientists. They are thus confined to thinking according to plan—at the expense of developing their own expertise and of achieving some grasp of various ways of standardizing the world (of which scientific standardization is but one variety; see also Aikenhead, this volume). In other words, they learn to be docile, "disciplined."

However, a completely different situation could obtain and, with this, another type of science education. Such education promotes a reflexive type of socialization in science, in particular by building on what students seem quite capable of doing when, precisely, they converse among themselves about "academic things."[1] That is, when they engage in "suspecting, pondering and putting the world to test" (Boltanski & Thévenot, 1991, p. 54; see also Roth and Bowen, this volume). Obviously, this type of learning

---

1. See the study by Fasulo, Girardet, and Pontecorvo (1998) conducted among primary-school children, who discussed the historical reliability of a photograph. See also the research by Driver, Leach, Millar, and Scott (1996), Larochelle and Désautels (2001), or Désautels and Roth (1999) among secondary-school students who discussed the world of science with their peers.

situation is much less sedate. However, in socioepistemological terms, it appears to hold out much more promise in that it affords students the opportunity to "practice social life" in an informed thinking way and thus gain awareness that neither scientists' nor their own knowledge and classifications are the result of spontaneous generation. In other words, a learning situation of the type I have referred to could offer them the opportunity to become aware that everything said is said by someone (to borrow Maturana's expression)—a dictum that obviously applies to my own comments.

In the next two sections, I will briefly outline the perspective that informs my reflection on both these questions and that of classification. Then, I examine the two cases of classification referred to above, giving particular emphasis to the case occurring in a school setting. In this classroom situation, we have a telling illustration of the invisible work of classification and of its inescapable political and ethical dimensions (Bowker & Star, 1999). Throughout this text, I articulate my contention that science education should recognize and work with these dimensions. It should, by the same token, reintegrate the voices of classifiers and references of the legitimization process. Such is a condition of advancing a type of science education that fosters social empowerment rather than discipline and docility.

On Pigeonholing Humans

Lucy: People aren't animals; they're humans.
Teacher: People are animals, the same as dogs and cats and so on.
[...]
Jimmy: But people talk, and have two legs and arms, and move and can think. Animals aren't like that.
(Laughter)
Teacher: [...] That's enough. People are animals.
                              (Excerpted from a science lesson at primary
                              school level, quoted in Munby, 1982, pp. 19–20)

For anyone interested in the potential socioepistemological problems involved in learning scientific knowledge and the power effects accompanying such knowledge (as Foucault [1971] would point out), the opening quotation offers an interesting situation. It provides an eloquent illustration of how, as early as primary school, the age-old distinction between "sacred" and "profane" knowledge is re-institutionalized. This re-institutionalization occurs via a format of interaction and a type of rhetoric that places the two registers of knowledge in competition with one another, to the advantage, obviously, of the "chosen" knowledge.

These children, who have only begun to take part in science education that so often trivializes profane knowledge, obviously do not submit to this curious classification at the outset. Such perspicacity can only be admired: If they had adopted this classification, they would have been forever barred access to all those places posting "No Animals," including even science classes! But it is easy to imagine that in fairly short order, if only to guarantee their survival in the school setting, they too will end up adopting this classification and expressing themselves according to the only publicly acceptable discursive code and content. In that connection, research into the responses of schoolchildren to insoluble or absurd problems is most instructive. For example, in one survey children intuited the absurdity of attempts to estimate the surface of a garden according to the number of rows of cabbages covering it (Schubauer-Leoni & Ntamakiliro, 1998). They nevertheless set to work as though they had indeed adopted the attitude of "academic correctness," a view which suggests that the problems presented by a person in a situation of institutional authority are necessarily relevant and that just as inevitably *a* response can be devised for them.[2] But in what way does such a situation touch on socioepistemological problems?

As with any problem, a socioepistemological problem is declared such not only on the basis of the model of cognition and science to which one subscribes but also according to the educational goals that one pursues. How a socioepistemological problem is formulated also depends on the sociopolitical project that

---

2. These responses might well testify to the representation that these children have made mathematics out to be. Belgian schoolchildren who were asked to comment on their responses to similar questions noted: "Sure it doesn't mean anything, but that's mathematics!" (Jacquard, 1986, p. 66).

more or less openly underlies one's actions and educational interventions (Fourez, 1992). For example, the views of Lucy and Jimmy indicate to epistemological realists problems or difficulties of comprehension stemming from either a preconception or a misconception. Thanks to a judiciously applied teaching approach that emphasizes unambiguous messages and activates the application and repetition of the knowledge being taught, such errors should be overcome or eradicated. In other words, this perspective rests on the assumption that the difficulty students have appropriating the codes and artifacts of scientific knowledge testifies to "a knowledge deficiency in need of repair" (Layton, 1993, p. 1). In short, a problem may be defined as that which diverges from the norm—a norm, moreover, which is assumed to exist in itself as opposed to existing by virtue of a convention or a standardization. This is the so-called cognitive deficit model under which the knowledge of students is deemed worthy of interest to the extent that the main issue is one of identifying what has gone wrong in relation to the "chosen" knowledge promoted in teaching curricula and textbooks.

It is quite a different story for those who envision learning from a constructivist perspective, thus with a view to "multiplying potentialities" (von Foerster, 1992). Thanks to an accumulation of experiences and deliberations, students develop viable narratives and explanations to deal with their day-to-day affairs (von Glasersfeld, 1995) and to participate in the customs of their membership groups. In so doing, they contribute to what Moscovici and Vignaux (1994) call "sociodiscursive cohesion," by means of which members of a group articulate, negotiate and standardize their relationship to the world (Larochelle, 1999). From an educational point of view, attention is focused on the potential for action contained in students' knowledge and the discursive resources that give this knowledge form and establish its relevance in relation to the context. Attention is also focused on additional possibilities afforded for identifying connections and distinctions with other action potentials and other ways of viewing and acting in the world, such as scientific conceptions. As with the preceding realist perspective, a process of comparison is most definitely brought into play but in view of a different set of objectives and in accordance with another representation of what education might mean. It is no longer a question of colonizing the knowledge of students with the knowledge of specialists and experts. Rather, it is a question of stimulating the informed appropriation of the customs, habits, alliances and productions of specialists/experts and of empowering each and every one to make a "judicious use of the experts" and their ways of standardizing the world (Fourez, 1994). In other words, it is a question of initiating students into another register of sociodiscursive cohesion and of seeing to it that this process preserves a reflexive character. This is done by "moving back and forth between life-worlds and the science world, switching language conventions explicitly, switching epistemologies explicitly" (Aikenhead, 1996, p. 41). Otherwise, we would be indulging in an idyllic interpretation of constructivism and, according to Muller and Taylor (1995), blurring the borders between these registers and trivializing the power relationship between them.

> [To] cross the line without knowing it is to be at the mercy of the power inscribed in the line.... This is not to say that we accept the line as legitimate, merely that the battle cannot be won by trying to erase it discursively.... [In other words] the progressive strategy consists in finding out how to empower people by ensuring that they have the wherewithal to cross the border safely. (p. 272)

This socioconstructivist perspective informs the remainder of my comments on the socioepistemological problems that may be entailed by the formats of science education, which, as was indicated in the quotation introducing this section, make scientific repertoires, classifications or distinctions a question of ontology. By the same token, these formats relegate other repertoires to a status that is less prestigious—that is, to the status of opinion and belief, in short, of repertoires that are "handicapped" by their sociocultural foundation and anchoring, and are therefore assumed to be contaminated by irrationality.

Classification: A Question of Ontology or Negotiation?

Think of "objectivity": it demands that the properties of the observer shall not enter into the description of his [or her] observations. I ask, how can this be done? Without him [or her] there would be no description nor any observation! (von Foerster, 1990, p. 10)

In traditional psychology, as in popular psychology, it is not unusual for concepts to become endowed with an ontology. That is, concepts are identified in terms of things, qualities, and forces belonging specifically to the person observed rather than to the observer's classifications, the connections s/he establishes, or the reigning *Zeitgeist*. The long and laborious career of the notion of "self" (as with "intelligence") provides ample proof of this tendency that imbues concepts with *"une épaisseur de réalité"* (a layer of reality, or the characteristics of a substance [Moscovici & Hewstone, 1984]). Thus, "self" has been frequently regarded as some sort of undeniable psychic entity or force, or as the center of a subjective consciousness having control over all our actions. In other words, according to this so-called realist perspective, "self" is "located" and "naturalized"—that is, ascribed to the nature of a person (Larochelle, 2000).

For the partisans of the (socio-)constructivist thesis, concepts are of an entirely different order, representing instead "keys which are useful to unlocking problems" (von Foerster, in Segal, 1986, p. 33). From this perspective, concepts do not fall within the scope of substances but rather of performances: The construction and maintenance of concepts depend on the performance that they give rise to over time and in space, and of course, within a particular sociocultural and sociolinguistic repertoire. Acknowledging the contingent character of concepts does not imply that the latter are thus deprived of robustness or sustainability. Nor does it rule out the possibility that they also end up becoming "naturalized" and endowed with a "layer of reality" to the point that they appear "to go without saying." Nevertheless, and this point forms a cornerstone of constructivism, there is no forgetting that this "ontology" is an artifact, an outcome, and not an a priori: it is *produced*.

In its own way, the classic Piagetian model led us into the same direction. It showed that the structuration of the world in young children essentially proceeds from the history of the success and failure of their schemes of action (i.e., what may be sucked, what may be stacked, what may be fitted into another thing) in relation to their goals. According to this view, by "computing," "putting together," and classifying their experiences, children establish lasting objects and a certain order of things that contributes to their impression of dealing with a "stable" world and, ultimately, of having a specific identity.

However, as shown in the above classroom episode, all classifying and categorizing activities do not contribute to creating an order of things as "tranquil" as that of the formal world of Piaget. Classifying and establishing the borders of an object within a whole can also *create hierarchy, or more exactly, opposition*" (Guillaumin, 1981, p. 33), and, at that point, give rise to particular social relationships. There are many historical examples that may serve as clear illustration of this social mechanism of differentiation by which the people subjected to it can be thingified. One such example is the taxonomy of "born criminals" developed by the late 19th-century Italian school (Darmon, 1989) or the cartography supposedly capable of accounting for the "mental troubles" of individuals (Rosenhan, 1988). Or again the scholastic categories and procedures by which students are declared gifted, "at risk," or "socially and emotionally maladapted" (Mehan, 1991), and the question of disciplinary territories (Messer-Davidow, Shumway, & Sylvan, 1993). Such differentiation makes it easier to handle, pigeonhole, and eventually exclude those that are thingified (e.g., Désautels, Garrison, & Fleury, 1998).

Obviously, it is not a question of falling for a conspiracy theory and concluding that all classification proceeds according to some Machiavellian design. Instead, my comments are intended as an invitation to view classification as the mouthpiece of those who have devised it and to take note of how, in terms of both its process and its products, classification is fraught with socioepistemological issues. It proceeds from decisions, is made possible by habits and patterns of thinking and speaking, and brings into play considerable stakes and tensions owing to the fact that all classification comes down to establishing the agenda of what is considered legitimate within a given society or group (Bowker & Star, 1999). The following case of classification is emblematic of the anchoring and the projects of the authors behind these efforts. Such anchoring and projects often become invisible once the classification in question has managed to become ingrained in either the theoretical domain or the popular imagination.

On the Pigeonholing of a Hybrid Object

During the 19th century, a debate raged among naturalists over the classification of a strange animal discovered in Australia and that today is known by the name of the duck-billed platypus (*Ornithorhynchus anatinus*). Standard dictionaries do not show an entry for "platypus" that would suggest what for some sociologists of science might be considered the "native environment" of this scientific object—that is, the tensions, conflicts, and negotiations that opened the way to its inclusion in the scientific order of things (Stengers, 1997). Much as with the discursive practices favored by science classes and textbooks (Lemke, 1993; Mathy, 1997; Sutton, 1998), ordinary dictionaries do not often resort to rhetoric of the kind "This is how this object was negotiated and established, along with the associated consequences and risks." Rather, they are much more partial to "bald objects" (Latour, 1999), that is, well-defined objects that are presented as though having emerged from nowhere, as though having no genitors, as though presenting no risks.[3] In short, they are objects that are presented in terms of "That's the way it is." The following quotation, from a widely-used dictionary of French, provides a clear illustra-

---

3. From Latour's point of view, a risk-free object may, of course, entail unintended consequences, hence may involve risks. However, these consequences are not viewed as being an integral part of the object in question but are instead conceived of as stemming from the impact of the encounter of this object and a different universe. An example of such an object is a meteor that collides with a social world to which it is foreign and which serves as a target to it. From this perspective, there is little likelihood that the consequences re-impact the definition or properties of this object: they stem from another order and are thus unattached to it.

tion of this tendency to short-circuit the networks, connections, and alliances that made it possible to *produce* the platypus and, in a way, socialize it within the world of scientists.

ORNITHORYNQUE (*platypus*) (masculine noun—1803; from *ornitho*–, and Gr. *runkhos*–beak) A semiaquatic oviparous mammal (of the monotreme family) having a corneous beak, a long flat tail, and webbed feet; found in Australia and Tasmania. (Robert, Rey, & Rey-Debove, 1996, p. 1550)

However, according to the accounts of Eco (2000), Deligeorges (1997) and Gould (1993), it would be more worthwhile conceiving of this animal in terms of the world of "disheveled objects." This is an allusion once more to a number of elements contained in Latour's comments—that is, fuzzily defined objects or objects that reveal a long series of connections and threaten to upset the hopes and projects assigned to them. Their genitors have not been hidden from view but are instead highly involved, committed if not controversial. In this connection, Eco's presentation (2000) of this strange creature is highly evocative, suggesting that it is rather unlikely that the classification of the platypus was as convivial an affair as is suggested by the dictionary. Even popular imagery, which nevertheless constitutes an almost inexhaustible repertory for reducing the hard-to-swallow easier–to–bite-size proportions, apparently was brought up short by this recalcitrant animal. Witness, for example, the expression "water-mole" with which the platypus was first associated. While being perceived as a kind of aquatic mole, it was a most surprising mole for having a bill. As Eco mentions, "something perceptible outside the 'mold' supplied by the idea of mole made the mold unsuitable" (p. 59) unless the mold was corrected and it was decided that the world now contained billed moles. Or again, unless yet another mold was brought forward, as suggested by the other terms used to refer to the platypus, such as "duck-mole."

The platypus is a strange animal. It seems to have been conceived to foil all classification, be it scientific or popular. On the average about fifty centimeters long and roughly two kilos in weight, its flat body is covered with a dark-brown coat; it has no neck and a tail like a beaver's; it has a duck's beak, bluish on top and pink or variegated beneath; it has no

outer ears, and the four feet have five webbed toes, but with claws; it
stays underwater (and eats there) enough to be considered a fish or an
amphibian. The female lays eggs but "breast-feeds" her young, even
though no nipples can be seen (the male's testicles cannot be seen either,
as they are internal). (Eco, 2000, p. 58)

The naturalists and taxonomists (British, German, etc.) who
examined the first stuffed specimens of the platypus apparently
had an impression similar to that of Lucy and Jimmy. They
thought they were confronted with something that foiled their
frames of reference and that indeed seemed to fall within the
scope of the unthinkable. Some suspected (yet) another hoax by
taxidermists, while others were of the opinion it would be better
to classify the creature under the heading of exceptions, while yet
others fell back on the categories then in use, among which those
of mammalian and oviparous animal were prominent. The scientific community became even more perplexed once specimens of
the male and female were made available (in alcohol) and the
practitioners of various disciplines could examine and (intellectu-
ally) dissect them at length. The codes of the era, the "disciplinary
divisions" and the established paradigms were all set on their
ears, so to speak: A furred animal that hatched from an egg? A
female with nipple-less mammary glands? How, then, could the
young possibly suckle when encumbered with such a bill? What,
then, was this beast that had the physiognomy of a chimera? A
bird, a fish, a mammal, a reptile?

Just as, following popular imagery, the front portion (or bill) of
the platypus stimulated scientific imagery, so too the rear portion
was by no means devoid of interest (Gould, 1993). The platypus
has a single orifice for all excretory and reproductive functions
(hence the term "monotreme"). Reproduction was, moreover, an
issue that sparked intense debate among partisans of the mammalian platypus option, those of the oviparous platypus option,
and finally, those of the ovoviviparous platypus option. The debate continued unabated in scientific circles and journals for more
than 80 years until 1884 (an era when, as everyone knows, Darwin's theory of evolution and the selection of species began to
subvert the paradigms of naturalists). At that point, W. H.
Caldwell (a scientist who had been conducting field research in

the "purest colonial style") sent the University of Sydney a tele-
gram that has remained celebrated in the annals of scientific history ever since. It read: "Monotremes oviparous, ovum meroblas-
tic," thus sealing the fate of the platypus—and scientific attempts
to coax the animal back into tried and true classifications. "Mero-
blastic" in fact referred to the mode of egg cell division, which, on
the basis of the yolk content, was revealed as being typically rep-
tilian (Gould, 1993). Thus the platypus turned out to be neither
mammalian nor oviparous but indeed both mammalian *and* oviparous.[4] So doing, the platypus joined the sparsely populated pantheon of the monotremes, in the company of the spiny anteater
(echidna), another Australian mammal that had successfully out-
foxed naturalists' efforts at classifying it. Even today, however, a
minority of scientists continue to consider monotremes as reptiles
of a certain type, and refuse to classify them under the heading of
mammals (Deligeorges, 1997).

On the Pigeonholing of Students

In a certain way, the history of the platypus bears a family resemblance to the classroom episode referred to above and that is reproduced in full below. In effect, this episode also displays an
object that turns out to be less "inert" than was initially imagined
and thus stands as a "kissing cousin" of the world of "disheveled" objects. The contours of the object "animals" are thus a bit
more blurred or at least a bit less reliable than first anticipated.
The protagonists bring into play a series of connections that themselves refer to different affinities and networks. All participants
are indeed highly involved, committed if not controversial. Thus,
the partisans of the oviparous platypus stressed the eggs and dis-
regarded the question of mammary glands while the partisans of
the mammalian platypus opted more readily for the mammary
glands and the milk and ignored the question of the eggs. Simi-

---

4. Obviously, the community in question here by no means has a monopoly over
platypus-type objects, considering the wave-particle duality of the concept of light
with which the community of physicists must contend with.

larly, the students advanced the repertoire of everyday things that excluded the academic blend of "people-animals" while the teacher advocated the academic repertoire and kept the everyday repertoire off limits. In other words, both scenarios offer an excellent illustration of how "observation sentences can be made only in the light of a conceptual framework or of a theory that gives them a sense" (Eco, 2000, p. 248).

However, this analogy cannot be pursued further, as the case of the platypus also offers an excellent illustration of how, in debate, the clash of resistances entailed by differing conceptual repertoires or affinities can indeed prove fertile. The partisans of a given thesis must "find" allies and evidence to convince opponents that what apparently passes for trifling concerns are indeed worthy of consideration. Furthermore, the opponents should account for this new evidence in their own theses. To return to the previous example, the female platypus does indeed lay eggs, but she also has mammary glands that are a delight to her young! In short, in scientific communities, everything takes place as though scientists became characters in one another's biographies (to paraphrase Geertz, 1999).

In the classroom episode, however, it is another story altogether. The affinities, repertoires, and disagreements at stake do not give rise to deliberation. The question of what might make it possible to delineate the difference between scientific and experience-based knowledge remains closed off from discussion. It is true that confrontations and resistances occur, as is shown in the original version of the episode presented below. However, as the criteria for success in this confrontation have been previously established ("People are animals"), such clashes as may occur lend themselves to negotiation but little. Only impacting or collision is an authorized mode of confrontation, meaning that the participants will be effectively prevented from becoming characters in each other's biographies, regardless of their status as students, teacher, or scientists:

Teacher: Now, we are going to leave the nonliving things for later and
           study just the living things. (Writes "living" on the board.) . . .
           Now, let's divide all the living things into two divisions. Into
           what two divisions can we divide every living thing? Every

living thing is either a _____ or a _____? Lucy, give me one division.

Lucy: People?

Teacher: People are just part of one of the two divisions.

Peter: Plants and animals.

Teacher: Good for you, Peter. That's right. Every living thing in this world is either plant or animal. People, Lucy, are animals, so they fit in this division.

Lucy: People aren't animals; they're humans.

Teacher: People are animals, the same as dogs and cats and so on. (Much laughter, and several loud objections by a large number of children speaking simultaneously. It appears that they disagree with this last statement.) People are *animals*. What's wrong with that? They're not plants, are they?

Jimmy: But people talk, and have two legs and arms, and move and can think. Animals aren't like that.

(Laughter)

Teacher: People do think, and this makes them one of the highest forms of animals, but they are still animals. . . . And other animals communicate with one another. (Several children are noisy and visibly disturbed.) . . .

Teacher: That's enough. People are animals. Now maybe it would help if we looked at the differences between plants and animals. What are the differences? There are at least three that you could name.

(Quoted in Munby, 1982, pp. 19–20)

This teacher thus attempted to impose what she felt was undoubtedly obvious, heedless of the usual but necessary precautions required in respect of the "disheveled" nature of the distinction involved. She thereby discredited the knowledge of these students (probably quite involuntarily). In addition, the initial elicitation (itself something of a prompting) which *appeared* to be concerned with what students know ("Lucy, give me one division") was abruptly and unequivocally shut off with the expected answer: "That's enough. People are animals." Exit the notion of the living and the problems of defining this concept both in the scientific sphere and in children's imagery (e.g., Laurendeau & Pinard, 1962; Piaget, 1972): classification is self-evident. At no time does the teacher bother discussing with the students how the categorization "person-animal-plant" may be quite viable and relevant in everyday life, but that in the academic sphere, on the other hand, it is (literally) another story. For various reasons, scientists have adopted another type of classification in accordance with other knowledge-related stakes and interests, even though, in everyday life, they too behave according to the criteria of Lucy and Jimmy. As this teacher's discourse would have it, classificatory criteria are not invented but are instead "visible" to any good observer. Nevertheless, if one adopts the students' point of view, is it not equally apparent that plants cannot walk? Is this fact not self-evident, too? In other words, the teacher has passed up an opportunity to initiate students into the diversity of knowledge games and the usefulness of moving from one game to another in a reasoned, informed way. Failure to do so might well have meant "mistaking one's audience" (Goffman, 1973).

The remainder of this episode is scarcely more promising. Despite her initial, mixed results, the teacher then announced that the class would pursue this task according to a similar type of approach: "Now maybe it would help if we looked at the differences between plants and animals. What are the differences? There are at least three that you could name." Thus, establishing a community of knowing was not the prime concern in this instance, nor was developing a form of reflexive comprehension of the object and the task at hand, nor, for that matter, was exploring the world of students and opening it to other potentialities. What counted was restating the declarations promoted in the school curricula and textbooks.

What, then, have these students learned? For a multitude of reasons, it is not unreasonable to think that they have "learned their place." Most likely, they have learned that two categories (plants and animals) suffice to classify living beings, and that these categories and these alone are acceptable, at least at school. By the same token, they have learned that the *knowledge that counts* is not their knowledge, hence that their experience-based knowledge is deficient and can be declared to be such in the absence of any argument or debate—by no means a trivial lesson. In other words, they most probably have learned that the traditional authority associated with social position suffices to establish *as truth* the knowledge which is taught at school and which claims to be derived from scientific knowledge. Then, to transfer this "state of grace" to sciences and their emissaries, there is but one step, easily taken and indeed encouraged by additional study of science, as is illustrated in this high school discussion concerning the controversy surrounding acid rain:

Student: It's like there are two different stories, the Canadians say it's from the States, and the States' guys say that you don't know where it's coming from.
Teacher: But . . . how do they get those two different conclusions based on the same observations?
Student: They're two different countries [Inaudible].
Teacher: *But they're scientists... and scientists should be people who look at the facts rather than look at the... you know, other things.* They should be basing it on the facts.

                                   (From Geddis, 1991, p. 179, my emphasis)

As we have discussed elsewhere (Larochelle & Désautels, 1997), there is reason to think that yet again, these students have (re-)learned that the *knowledge that counts* has little to do with the knowledge they have developed. After all, no one has asked them how they managed to become such clever sociologists that they successfully scented a possible connection between the political game and the views put forward by scientists. Students may well have learned that actively taking part in the enterprise of knowledge that counts means (as Rorty [1990] put it cynically) permanently exposing themselves to "hard reality." As part of this process, they will must disregard themselves, their beliefs, commitments and affinities for the community of scholars is reserved for "people who look at the facts rather than look at . . . you know, other things." It is equally likely that they are aware that by being in contact with things outside the realm of human relations, the members of this community have also escaped all form of human accountability—a lesson that, like the previous episode, is not trivial lesson either. To entertain such considerations is far from indulging in pure speculation as research among secondary school students has shown (Driver *et al.*, 1996; Ryan & Aikenhead, 1992). The following comments of a high school senior are rather representative of those current among the population concerned as we have shown elsewhere (Désautels & Larochelle,

1998); it portrays the tendency to view scientists as mere "observ-ers standing outside the world and time" (Mathy & Fourez, 1991):

> It seems to me, when something's scientific, it can only be objective. Oth-erwise, it would be subjective, and then it would be irrelevant, I think. *Your [personal] ideas don't have anything to do with it,* because science is something that is precise. You don't interfere with it. . . . You can try to understand [phenomena] and master them. You can't really, you can't change them really. *You can't get involved in them.* (Désautels & Laro-chelle, 1989, p. 96, my emphasis)

It may be obviously objected that even within a context of authority as firmly instituted as that in which the referred-to epi-sodes unfolded, students can all the same develop a more eman-cipatory relationship toward science than that suggested by my comments. At the outset, they could develop a certain skepticism toward the "epistemological exceptionalism" (Bimber & Guston, 1995) that currently halos the scientific enterprise inside the class-room. Undoubtedly, such an alternative scenario is possible, but it should not cause us to ignore the fact that the disciplinary power underlying teaching practice represents a discipline of both body and mind.[5] Furthermore, as Roth and McGinn (1998) have brought out with clarity, this discipline, like disheveled objects, is en-dowed with far-reaching connections and barely graspable ramifi-cations, while at the same time being devoid of authors, in the manner of "bald" objects. In other words, disciplinary power is a formidable but discrete power (Foucault, 1975). Invisibility is key to producing its effects, but on the other hand, it constrains all who are governed by it to a principle of mandatory visibility serving to maintain them in a state of subjection, and making the student's job a most peculiar one indeed:

---

5. In that connection, the story of the platypus is again instructive. The protago-nists indeed appear to be "disciplined" by their discipline, their own categories or paradigmatic divisions. They are disciplined to the extent that whenever they at-tempted to imagine a different scenario for "reining in" this animal, they could do no better than combine the previously existing divisions or classifications to form the category of "mammalian-oviparous" (e.g., Messer-Davidow, Shumway, & Syl-van, 1993).

> If a student's job is a most peculiar one, this is not, in the first place, be-
> cause it is unpaid. It is because it: is not chosen freely, or less so than any
> other; is heavily dependent on a third party, not only in terms of its main
> finalities and conditions, but also of its details, and particularly of its
> fragmentation and relationship to time; is performed under the gaze and
> control of a third party, not only in terms of its results, but also of its
> slightest procedures; and provides a continual basis for evaluating the
> qualities and defects of a person, his/her intelligence, culture and char-
> acter. Some adult jobs are just as restrictive (forced labor, prostitution)
> as that of being a student. Others are just as dependent (the most un-
> skilled jobs). A number of jobs are closely monitored by others or are at
> least subject to being watched, while others involve judgments on indi-
> viduals. But only rarely are all these features combined [as in the case of
> students]. (Perrenoud, 1995, p. 14)

It should not be forgotten that this power does not merely re-
press, obscure, and censure: in a way, it also makes things possible (Foucault, 1975). It in fact *produces*, that is, establishes what
should be held for admissible and reasonable. Thus, in a phenomenon deserving of greater scrutiny, the school of science seems
to have an impact on the ways students "design their future,"
even when they turn away from it. The following point of view
provides an eloquent illustration of the taboos participating in its
author's design of her future. In many respects, it is consistent
with the view that mathematics and science teaching constitute a
source of symbolic violence for many people in our societies (Trabal, 1997).

> I wanted to be optometrist. That's why I took physics. I used to be good
> at it. I always had high grades in science. Now, I just don't understand
> physics, I can't do physics anymore. I used to be able to, but I can't any-
> more. No matter how much I study. On the weekend we had a chemistry
> quick quiz and no matter how much I studied or wrote down and read
> through the book about three times, I got bad marks. I just can't do sci-
> ence. I don't want to become an optometrist anymore. Physics was just
> too hard. I got poor grades. Now I am just doing it as a requirement to
> complete the number of courses I need to finish school. (High school phys-
> ics student, quoted in Roth & McGinn, 1998, p. 400)

To wit, there is room to think that at the school of science,
students learn a great deal more than a set of formulae, laws,
theories—and facts. They also develop a certain repertory of be-

liefs about science and its finalities, a certain discursive register
with which to discuss science and the way it is conducted, and, of
course, a certain relationship to the world of knowledge, which,
more often than not, is fundamentally asymmetric in nature. On
the one hand, there is the (scientific) knowledge of *the knowers*,
and, on the other, the (profane) knowledge of *the believers*.

Likewise, science teaching can scarcely be viewed as simply a
question of "passing on" formulae, laws, theories—and facts. Inevitably, as the above-mentioned classroom episodes go to show,
science teaching comes down to choosing from a range of narratives about science. Such a choice obviously proceeds according to
the representation of science one endorses and maintains with respect to the competence and expertise of the people for whom this
teaching has been intended. However, it is no less clear that such a
choice is also determined by the relationship to knowledge and
power that one holds to be desirable and that refers to the social
project that one *wants* to promote. In other words, science teaching
amounts to the teaching of a sociopolitical and ethical project. The
"whys and wherefores" of these projects obviously vary depending on whether emphasis is placed on the tranquil world of
"That's the way it is" or the disheveled world of "That's the way
it was negotiated and established." Whatever route one opts for,
it is quite clear that students learn, design their future, and hence,
practice in social life. All the same, it stands to reason that not all
routes are productive of the same action potential nor generate the
same degree of intelligibility with respect to the world of scientists
and the controversies that are coterminous with that world.

In a certain way, the classroom episode involving the controversy surrounding acid rain again offers much useful insight. By
supporting his position with a world of bald objects ("the facts"),
the teacher is unable to capitalize on the disagreement outstanding—that is, make it intelligible. He can only save the existing or-
thodoxy—and facts—and thus plunge headlong into a teaching
impasse. Not only is he backed into relegating the disagreement to
the realm of the unthinkable, he also cuts himself off from establishing some sort of bridge between what students know and what
he would like to introduce them to. On the other hand, the students intuitively make the disagreement "manageable," so to

speak, by locating it from the outset in a world they are familiar with—a world in which the actors are involved, committed, controversial, much as in the everyday world.

In other words, by *re-socializing* the world of bald objects the way these students have (that is, by connecting it to the interests, projects, and networks this world is the product of), the result is an entirely different relationship to knowledge, an entirely different educational situation, not to mention an entirely different way of practicing social life. This is so, for the inevitable confrontation entailed by all educational action no longer opposes a group of subjects on the one hand *and* a world of finished, anonymous objects on the other. Rather, such confrontation emerges, as in the case of the platypus, as an encounter *between* groups of actors having different types of historicity. Or again the confrontation emerges *between* groups of "world describers" (students, scientists, teachers) who can deploy their "characters," ideas and connections. At that point, they are able to create the necessary conditions to discuss, negotiate, and hybridize the descriptions and worlds at issue, and to learn "how to make good use of these descriptions and worlds."

Conclusion

In this chapter, I articulate the socioepistemological problems that may result from unreflexive socialization into the world of scientific knowledge and classifications. I showed how it may be of particular interest to examine the usual confrontation that inevitably underlies formal educational practice and, more particularly, the practice of science education. Indeed, most of the time, this confrontation is consistently biased toward scientific classifications and distinctions, and endows these with a kind of epistemological exceptionalism—hence, constitutes them as a system of truth that would appear to lie beyond the reach of profane classifications and distinctions. So doing, this confrontation more closely resembles a case of domestication and disciplining than an opportunity for all participants to emancipate themselves by means of opening themselves, reflexively and critically, to other ways of standardizing the world, and of learning how to "make good use of this standardization."

Nevertheless, framing an event, a phenomenon or a situation according to a single theoretical referent is rarely fertile, even in the world of scientists, as the story of the platypus goes to show. The recent case of "Dolly" is also instructive in that respect. No doubt, the weirdness of pigeonholing and associating this clone to a single referent would be even more striking in the hands of Borges. For what is Dolly, actually? Can she be viewed in terms of a classroom distinction as only an "animal"? Or is she not also a major economic phenomenon on account of the future she has laid open to pharmaceutical firms? Is Dolly not also a phenomenon whose connections (and offspring) risk upsetting the traditional kinship structures: In other words, is the clone of our uncle our uncle? And, at that point, does Dolly not become prize game for lawyers in claims over inheritance?

The question arises of how children might venture to tinker and piece together a hybrid classification for this type of object? I am thinking of the same children who, beginning in primary school and throughout the remainder of their education, learn that science reveals the ontological order of things. These children also learn that only scientific distinctions are legitimate and authorize only a single type of relationship between a "content" and a "container" and that what they know or believe is inadmissible and that they have every interest in remaining within the paradigm of spectator. How might they authorize themselves to question the public discourse of scientific experts who make Dolly a "risk-free material" or a "bald" object? In short, how might they authorize themselves to take part in the trial of such a disheveled object as this quadruped that bleats, capers, and reaps royalties? How might students become participants in such discourse in the absence of an education that encourages them to be legitimate interlocutors capable of "suspecting, pondering, and putting the world to test" (including the world of scientists)? All in all, is putting the world to test not an inescapable condition of performing science as the case of the platypus would have us believe?

Acknowledgments

In this chapter, which was made possible in part thanks to a grant from the Social Sciences and Humanities Research Council of Canada, I further develop a number of elements that were previously dealt with in two articles written in French (see Larochelle, 1998; Larochelle & Désautels, 1997). I also wish to thank Donald Kellough, the translator of this text, for his fine work.

References

Aikenhead, G. S. (1996). Science education: Border crossing into the subculture of science. *Studies in Science Education, 27*, 1–52.

Bimber, B., & Guston, D. H. (1995). Politics by the same means: Government and science in the United States. In S. Jasanoff, G. E. Markle, J. C. Petersen, & T. Pinch (Eds.), *Handbook of science and technology studies* (pp. 554–571). Thousand Oaks, CA: Sage.

Boltanski, L., & Thévenot, L. (1991). *De la justification; les économies de la grandeur* [On justification: "Economies" and states of greatness]. Paris: Gallimard.

Bowker, G. C., & Star, S. L. (1999). *Sorting things out. Classification and its consequences.* Cambridge, MA: MIT Press.

Darmon, P. (1989). *Médecins et assassins à la Belle Époque* [Physicians and assassins during the Belle Époque]. Paris: Seuil.

Deligeorges, S. (1997). Puzzle sur pattes à bec et poils [A beaked, four-legged, furred puzzle]. *La Recherche, 296*, 62–65.

Désautels, J., Garrison, J., & Fleury S. C. (1998). Critical-constructivism and the sociopolitical agenda. In M. Larochelle, N. Bednarz, & J. Garrison (Eds.), *Constructivism and education* (pp. 253–270). New York: Cambridge University Press.

Désautels, J., & Larochelle, M. (1989). *Qu'est-ce que le savoir scientifique? Points de vue d'adolescents et d'adolescentes* [What is scientific knowledge? Some views of adolescents]. Québec, QC: Presses de l'Université Laval.

Désautels, J., & Larochelle, M. (1998). The epistemology of students: The 'thingified' nature of scientific knowledge. In B. J.

Fraser & K. Tobin (Eds.), *International handbook of science education* (pp. 115–126). Dordrecht, Netherlands: Kluwer Academic Publishers.

Désautels, J., & Roth, W.-M. (1999). Demystifying epistemological practice. *Cybernetics & Human Knowing, 6*(1), 33–45.

Driver, R., Leach, J., Millar, R., & Scott, P. (1996). *Young people's images of science.* Buckingham, England: Open University Press.

Eco, U. (2000). *Kant and the platypus. Essays on language and cognition* (A. McEwen, trans.). New York: Harcourt Brace & Company.

Fasulo, A., Girardet, H., & Pontecorvo, C. (1998). Historical practices in school through photographical reconstruction. *Mind, Culture, and Activity, 5,* 253–271.

Foucault, M. (1966). *Les mots et les choses. Une archéologie des sciences humaines* [The order of things: An archaeology of the human sciences]. Paris: Gallimard.

Foucault, M. (1971). *L'ordre du discours* [The discourse on language]. Paris: Gallimard.

Foucault, M. (1975). *Surveiller et punir. Naissance de la prison* [Discipline and punish: The birth of the prison]. Paris: Gallimard.

Fourez, G. (1992). *Éduquer. Écoles, éthiques, sociétés* [Educating. School, ethics, and society]. Brussels: De Boeck.

Fourez, G. (1994). Des objectifs opérationnels pour l'A.S.T. et îlots de rationalité [Operational goals for scientific and technological literacy, and islands of rationality]. In G. Fourez (with the coll. of V. Englebert-Lecomte, D. Grootaers, P. Mathy, & F. Tilman), *L'alphabétisation scientifique et technique. Essai sur les finalités de l'enseignement des sciences* (pp. 49–67). Brussels: De Boeck.

Geddis, A. N. (1991). Improving the quality of science classroom discourse on controversial issues. *Science Education, 75,* 169–183.

Geertz, C. (1999). *Savoir local, savoir global. Les lieux du savoir* (D. Paulme, trans.) (2nd edition) [Local knowledge. Further essays in interpretive anthropology]. Paris: Presses universitaires de France.

Goffman, E. (1973). *La mise en scène de la vie quotidienne* (A. Accardo, trans.) [The presentation of self in everyday life]. Paris: Minuit.

Gould, S. J. (1993). Qu'est-ce qu'un ornithorynque? [What is a platypus?]. In *La foire aux dinosaures. Réflexions sur l'histoire naturelle* (pp. 248–58) (M. Blanc, trans.). Paris: Seuil.

Guillaumin, C. (1981). Le chou et le moteur à deux temps: De la catégorie à la hiérarchie [The cabbage and the two-cycle engine: From category to hierarchy]. *Le Genre Humain, 2,* 30–36.

Jacquard, A. (1986). Deux et deux ne font pas quatre [Two plus two do not equal four]. *Le Genre Humain, 14,* 63–67.

Larochelle, M. (1998). La tentation de la classification. . . . Ou comment un apprentissage non réflexif des savoirs scientifiques peut donner lieu à un problème épistémologique [The siren song of classification. . . . Or how unreflexive science learning can generate epistemological problems]. *Recherches en Soins Infirmiers, 52,* 72–80.

Larochelle, M. (Ed.). (1999). Radical constructivism at work in education: An aperçu. *Cybernetics & Human Knowing, 6*(1), 5–7.

Larochelle, M. (2000). Radical constructivism: Notes on viability, ethics and other educational issues. In L. P. Steffe & P. Thompson (Eds.), *Radical constructivism in action: Building on the pioneering work of Ernst von Glasersfeld* (pp. 55–68). London: Falmer Press.

Larochelle, M. & Désautels, J. (1997). L'éducation aux sciences: l'"effet Fourez" [The "Fourez effect" in science education]. *Revue des Questions Scientifiques, 168,* 347–358.

Larochelle, M., & Désautels, J. (2001). Les enjeux socioéthiques des désaccords entre scientifiques: un aperçu de la construction discursive d'étudiants et étudiantes. *Canadian Journal of Science, Mathematics and Technology Education/Revue canadienne de l'enseignement des sciences, des mathématiques, et de la technologie, 1,* 39–60.

Latour, B. (1999). *Politiques de la nature. Comment faire entrer les sciences en démocratie* [Politics of nature: How to bring science into democracy]. Paris: La Découverte.

Laurendeau, M., & Pinard, A. (1962). *La pensée causale* [Causal thinking in the child]. Paris: Presses Universitaires de France.

Layton, D. (1993, October). *Inarticulate science? Or, does science understand its public as well as its public understands science?* Paper based on address given at the 3rd Royal Bank Lecture, hosted by the Mathematics, Science and Technology Group, Faculty of Education, Queen's University, Ontario, Canada.

Lemke, J. L. (1993). *Talking science. Language, learning and values.* Norwood, NJ: Ablex.

Lévy-Leblond, J.-M. (1984). *L'esprit de sel. Science, Culture, Politique.* Paris: Seuil.

Mathy, P. (1997). *Donner du sens aux cours de sciences. Des outils pour la formation éthique et épistémologique des enseignants* [Making sense of the science curriculum: Tools for the ethical and epistemological education of teachers]. Brussels: De Boeck.

Mathy, P., & Fourez, G. (1991). *Enseignement des sciences, éthique et société* [Science teaching, ethics, and society]. Namur, Belgium: Facultés Universitaires de Namur, Département "Sciences, Philosophies, Sociétés."

Mehan, H. (1991). The school's work of sorting students. In D. Boden & D. H. Zimmerman (Eds.), *Talk and social structure* (pp. 71–90). Berkeley: University of California Press.

Messer-Davidow, E., Shumway, D. R., & Sylvan, D. J. (Eds.). (1993). *Knowledges: Historical and critical studies in disciplinarity.* Charlottesville: University Press of Virginia.

Millar, R., & Wynne, B. (1988). Public understanding of science: From contents to processes. *International Journal of Science Education, 10,* 388–398.

Moscovici, S., & Hewstone, M. (1984). De la science au sens commun [From science to common sense]. In S. Moscovici (Ed.), *Psychologie sociale* (pp. 539–566). Paris: Presses Universitaires de France.

Moscovici, S., & Vignaux, G. (1994). Le concept de thêmata [On the concept of themata]. In Ch. Guimelli (Ed.), *Structures et transformations des représentations sociales* (pp. 25–72). Lausanne, Switzerland: Delachaux & Niestlé.

Muller, J., & Taylor, N. (1995). Schooling and everyday life: Knowledges sacred and profane. *Social Epistemology, 9,* 257–275.

Munby, H. (1982). *What is scientific thinking? A discussion paper.* Ottawa: Science Council of Canada.

Perrenoud, P. (1995). *Métier d'élève et sens du travail scolaire* (2nd edition) [The job of student and the meaning of school work]. Paris: ESF Éditeur.

Piaget, J. (1972). *La représentation du monde chez l'enfant* [The child's conception of the world]. Paris: Presses universitaires de France.

Robert, P., Rey, A., & Rey-Debove, J. (1996). *Le nouveau Petit Robert: dictionnaire alphabétique et analogique de la langue française* (2nd edition) [The new Petit Robert: Alphabetical and analogical dictionary of French]. Paris: Dictionnaires Le Robert.

Rorty, R. (1990). La science comme solidarité [Science as solidarity]. In *Science et solidarité* (pp. 46–62) (J.-P. Cometti, trans.). Paris: L'Éclat.

Rosenhan, D. L. (1988). Etre sain dans un environnement malade [On being sane in insane places]. In P. Watzlawick (Ed.), *L'invention de la réalité. Contributions au constructivisme* (pp. 131–160) (A.-L. Hacker, trans.). Paris: Seuil.

Roth, W.-M., & McGinn, M. K. (1998). >unDelete science education: /lives/work/voices. *Journal of Research in Science Teaching, 35,* 399–421.

Roth, W.-M., & Roychoudhury, A. (1993). The nature of scientific knowledge, knowing and learning: The perspectives of four physics students. *International Journal of Science Education, 15,* 27–44.

Ryan, A. G., & Aikenhead, G. S. (1992). Students' preconceptions about the epistemology of science. *Science Education, 76,* 559–580.

Schubauer-Leoni, L., & Ntamakiliro, L. (1998). The construction of answers to insoluble problems. In M. Larochelle, N. Bednarz, & J. Garrison (Eds.), *Constructivism and education* (pp. 81–103). New York: Cambridge University Press

Segal, L. (1986). *The dream of reality. Heinz von Foerster's constructivism.* New York: Norton.

Stengers, I. (1997). *Sciences et pouvoirs. Faut-il en avoir peur?* [Science and power: Is there cause for fear?]. Brussels: Labor.

Sutton, C. (1998). New perspectives on language in science. In B. J. Fraser & K. Tobin (Eds.), *International handbook of science education* (pp. 27–38). Dordrecht, Netherlands: Kluwer Academic Publishers.

Trabal, P. (1997). *La violence de l'enseignement des mathématiques et des sciences* [Mathematics and science teaching as a source of violence]. Paris: L'Harmattan.

von Foerster, H. (1990, April). *Lethology: A theory of learning and knowing vis-à-vis undeterminables, undecidables, unknowables.* Paper presented at the International Seminar "Conoscenza come educazione," Trento, Italy.

von Foerster, H. (1992). Ethics and second-order cybernetics. *Cybernetics & Human Knowing, 1,* 9–19.

von Glasersfeld, E. (1995). *Radical constructivism: A way of knowing and learning.* London: Falmer Press.

The Enactment of Epistemological Practice as Subversive Social Action, the Provocation of Power, and Anti-modernism

Jacques Désautels, Stephen C. Fleury, & Jim Garrison

This chapter is a continuation of our earlier work, which had us develop a version of social constructivism in education we called "critical-constructivism" (Désautels, Garrison, & Fleury, 1998). In the first section, we summarize some of the conclusions we reached that are important to make sense of the reflection to which we devote ourselves here. Then, drawing on episodes from a variety of school settings, we illustrate the fact that children are able to problematize the production and distribution of scientific knowledge that enables them to question, and, eventually, confront the arbitrary distribution of power in society. In other words, we show how enacting epistemological practice may be viewed as a powerful tool to challenge the traditional relationship to knowledge and power allowing students to make the first steps in an emancipatory journey. Finally, we briefly outline how aesthetics is inescapably part and parcel of the knowledge production as well as of students' ways of appropriating knowledge. Accordingly, we suggest that to become a liberating form of education, science education has to take into consideration aesthetic values.

On Critical-Constructivism

Our position runs counter to the received dogma whereby the disciplines (mathematics, biology, etc.) are assimilated to notional and methodological contents having an existence of their own materialized in books or other artifacts. Critical constructivists assert that all knowledge is the product of social processes involving a myriad of associated human and nonhuman actors (Latour, 1999a). More so, constructivists think that it is a terrible mistake to confuse the contingent products with the conditions that prevailed before they began. When this confusion occurs, few can resist the temptation to reify as objectively necessary what is actually a contingent product. This error resembles confusing wine in a bottle with grapes in a vineyard. Since knowledge is the product of social constructive processes and since we may fabricate products many ways (just as we may produce different kinds of wine from the same grape), all knowledge is contingent, and all scientific judgements are continually open to reconstruction and deconstruction.

These sociohistorical processes can be assimilated to what Foucault (1971) calls power/knowledge which produces not only contingent "objects of knowledge" and "rituals of truth" but also a social disciplinary order (Messer-Davidov, Shumway, & Sylvan, 1993). Thus, at any moment in the historical time of a society, only certain discourses will enjoy a high degree of legitimacy and be considered as valid knowledge instead of mere belief (Barnes, 1990). The social prestige associated with a particular type of discourse varies according to its location in the arbitrary hierarchy of knowledge promoted in a society, thereby generating an uneven distribution of sociopolitical power among citizens. In other words, the criteria according to which the truthfulness of different statements will be judged—dictating the "truth about the truth" (Delbos, 1993)—are relative to the social conditions under which they are produced. When the disciplinary order is contested, the reactions of the different stakeholders can be quite robust.

The historian of science Mario Biagioli (1993) exemplifies the position outlined above about the production of knowledge. He bases his argument on two tenets. First, during seventeenth century Italy, mathematical knowledge was considered a mechanical art or a technical knowledge that was mostly learned through apprenticeship. Second, it occupied a low echelon in the hierarchy of the disciplines in the university. In the following excerpt, Biagioli (1993) gives a description of the power/knowledge process in action.

> Copernicus and some of his followers faced a crucial obstacle when they tried to legitimize their work as not only a mathematical computational model but also a physical [or real] representation of the cosmos. The received hierarchy among the liberal arts was such an obstacle. According to this hierarchy (one that was justified by scholastic views on the difference between the disciplines and their methodologies), mathematics was subordinated to philosophy or theology. The mathematicians were not expected (or supposed) to deal with the physical dimensions of natural phenomenon, which (together with the causes of change and motion) were considered to be the philosophers' domain. Consequently, the philosophers perceived Copernicus not just as putting forward a new planetary theory, but as "invading" their own disciplinary and professional domain. In general, this invasion was unacceptable to them and, having higher disciplinary status than the mathematicians, the philosophers had resources to control such an invasion. The usual tactic (one that worked quite well in institutions that accepted this disciplinary hierarchy) was to delegitimize the mathematicians' claims by presenting them as coming from a lower discipline. (pp. 5–6)

We can then understand why Cardinal Bellarmine—in view of a possible confrontation with the theologians of the Roman Catholic Church—advised Galileo to disclose his thesis as an hypothesis and not as a description of reality, a confrontation that resulted as we all know in Galileo's condemnation, an episode of history that dramatically shows how the production of knowledge can never be dissociated from disciplinary power and its inscription in society. Engaging children of all ages in epistemological practices therefore constitutes a form of subversive social action (on this point, see also that chapter by Roth and Bowen, this volume).

In our earlier work, we also contended that by dismantling truth as a form of correspondence to reality and recognizing that there are no ultimate foundations of knowledge, critical constructivism "can have a potential emancipatory effect in education"

(Désautels et al., 1998, p. 255). To realize this potential we endorsed "the ethical necessity of applying the principle of epistemological symmetry—that is, symmetry of a type which considers the knowledge developed by students in the context of their local culture as viable and genuine" (p. 255). School knowledge is considered but one of the instruments in helping students to emancipate themselves from their own biographies. In this sense, reconstructing biographies in liberating ways requires participating in new social practices so that new meanings and forms of knowledge can be appropriated without negating one's own cultural knowledge.

The realization of the emancipatory potential of critical constructivism requires helping students "transform radically their rapport au savoir (relationship to knowledge)" (p. 256). We conceive of this as "a relation of meaning, and thus of value, between an individual (or group) and the processes or products of knowledge production" (Charlot, Bautier & Rochex, 1992, p. 29). If students cannot bring all they know, including their ways of knowing, their epistemologies, to the use and evaluation of scientific knowledge, then they will develop an inhibitory or disempowering relation to knowledge. Here we come face to face with questions of norms and values: issues of power become again salient. Value conflicts are always power conflicts. Whose values, whose norms, should prevail? That of the students, the parents, the community, or only those of the sciences? Science prevails when it constrains value consideration to epistemic values alone, and then only to its own epistemological norms. Once we see that the sciences and science education must consider aesthetic and ethical values, and not just epistemic values, the social sphere in which power operates becomes unrestricted.

Enacting Epistemological Practice in the Classroom

Before analyzing several student conversations, we clarify our understanding of epistemological practice. Epistemology as a discourse about the conditions of the production and circulation of knowledge is the result of socially situated practices. Epistemology, therefore, is conceived as the product of practices and not as a theoretically driven mental process enacted only by great minds. This is not to say that we cannot distinguish between neophyte and professional epistemologists who are aware of particular traditions in their field. Rather, one can do epistemology from a very young age. Drawing on the work of Fasulo, Girardet, and Pontecorvo (1998), we illustrate how nine- and ten-year-old children engage in such a practice as part of a history course. Following this illustration, we analyze the discourses of older students reflecting on the epistemological status of scientific knowledge.

In an Italian elementary school, ten groups of 4 or 5 fourth grade students were asked to make an interpretation of a photograph. The children had no previous experience analyzing visual documents and worked simultaneously for 40 minutes. Free discussion was elicited through the following task: Every group was given a photograph illustrating the inside of a Viking house and a sheet of paper with the following instructions:

> This is a reconstruction of a Viking dwelling (a people who lived in Northern Europe). Imagine we don't know anything about these people and this house is all that is left. Observe this photograph carefully and discuss it with your group mates trying to answer the questions. Try to reach an agreement and then write, in the first column, the things that are likely, and in the second, the things you are uncertain about.
> 1. From what you see, where did they cook their food?
> 2. From what you see, what did they wear and how did they defend themselves from cold?
> 3. From what you see, what kind of work were they doing?
> 4. From what you see, what kind of things were they able to make?
> 5. From this photograph, what other information can you get about the Vikings?
>
> (Fasulo et al., 1998, p. 258)

The children had to write down their answers in one of the two columns as "certain things" or "uncertain things" that generated some confusion among them because the instructions included the expression "likely" instead of "certain." The discussions were taperecorded, which enabled our subsequent analysis showing them enacting epistemological practice. The children tackled a variety of questions, including the one about the type of information that can be obtained from examining a photograph closely.

Before examining an excerpt of the children's discourses, one has to know that earlier in the conversation they challenged the idea that the Vikings could build houses at all, hence the problem of determining how some of them could own one. During that part of the conversation, Fulvio held that it was likely that it could have been exchanged for something else. Marco saw this possibility as being doubtful. Giovanni claimed that it should be classified as uncertain because "you can't know what is true, what Marco or Fulvio says, at least we aren't able to understand not even from the picture" (p. 263). So the excerpts presented below can be interpreted as a discussion between three young boys about truth and understanding, definitively classical epistemological topics. Let us then try to follow them in their course of action.

> Giovanni (1): Actually to me Marco is right because how can one know whether that house has been conquered, remade, modified because you see it as it is now in the present. You can't see it as it is in the future [probably read past].
>
> Fulvio (1): All right, but it is still likely [that the house is there because it has been exchanged for something else] because we haven't a document to tell us about the past.
>
> Marco (1): Well then, it's a discourse that cannot end because either someone gives us some document that gives information on the other times, or it's a discourse that cannot end.
>
> Giovanni (2): To me it is as Marco says, that this is a discourse that one cannot end because we don't know enough things. This photograph doesn't tell us enough things about where to put it, about what we can know precisely.
>
> Fulvio (2): To me we need to find a solution but you say, Marco, that I have to bring another document, then you have to.
>
> Marco (2): No.
>
> Fulvio (3): What can we do then? This document is somewhat clear, you can see the fire. It's clear this document; it just doesn't fit the discourse we are doing. (p. 263)

Clearly enough, these children do not appear newcomers to the art of argumentation. Our main interest lies with the epistemological dimensions involved in their linguistic interaction. We can distinguish at least three argumentative moves in Giovanni's (1) first intervention to counter Fulvio's claim that the statement about the possibility that the house was exchanged for something else should enter into the category labeled "likely." He first asserts that Marco's point of view is right, which can be interpreted as a way to mobilize an ally for his cause. Then, using a strategy of particularization, which is rhetorically quite efficient in such circumstances (Billig, 1996), he claims that the process of exchange alluded to by Fulvio is simply a particular case among a panoply of possibilities (conquest, remodeling, modification, etc.). Through this move, he tries to shift the focus of the conversation from the problem of categorizing one particular statement (the exchange) to the specification of the conditions for making a judgement about such inferences. He finally draws on a fresh epistemological resource to complete his argumentation. Indeed, as he argues, all of those possibilities have to be classified in the uncertain category since one cannot infer from a photograph which one was enacted since "you see it as it is now." In so doing, Giovanni seems to hold that it is only through access to past events that a decision can be made and lifted. Historical time then becomes a potential object of discussion.

Fulvio (1) mounts a counter-argument and tries to return to the topic of discussion so as to argue about the categorization at hand (likely and uncertain). First, he acknowledges that it is necessary to have recourse to past events through the consultation of documents but holds that one can still classify the statement in the category of likely events. In other words, while agreeing with Giovanni about the necessity to frame the events in historical time, he tries to bring the focus on the categories themselves, but apparently without much success.

Marco (1) ignores the fact that his two peers seem to be using different categories (likely and certain) to make their judgement about the different hypothesis relative to the presence of the house. He restates the necessity to consult documents about the past to make a decision and concludes that otherwise the discussion could go on forever. In his next intervention, after acknowledging Marco's point of view, Giovanni (2) makes clear that the photograph in itself cannot be of any help to make a decision about the certainty of the statements. In his words, it is not an instrument that can tell them what they can know precisely.

The rest of the conversation takes another turn. Fulvio's (3) intervention is directly addressed to Marco in relation with the necessity to use documents. If he, Fulvio, has to bring a document then Marco has to do the same. But Marco's response is quite clear: he simply replies with a straightforward "No," a response that can be interpreted in many ways. It could simply mean that if Fulvio brings a document, then the others do not have to do the same because one source is enough. It could also mean that Marco was aware of Fulvio's (1) past rhetorical move and estimated that one does not have to use a document in order to decide that statements are uncertain from just looking at a photograph. Such decisions are much more difficult to make for the category entitled "likely."

Finally, Fulvio (3) seems to surrender to his peers' arguments and makes quite clear that they, as a group, had entered into what can be interpreted as an epistemological dilemma. It is possible to recognize some familiar objects in the photograph (fire, pottery, etc.), but there is a limit to the knowledge that can be obtained from such an operation. Moreover, this artifact, as he says "just doesn't fit the discourse we are doing." In other words, some other instruments are needed to expand their knowledge about what the Vikings could do or not do!

We contend that through their argumentative business, these young children enact an epistemological practice. They tried to circumscribe the necessary conditions that have to be obtained in order to classify different statements related to historical events as either certain or uncertain and concluded that the photograph cannot provide them with the proper information since it says nothing about the past, it is timeless. In other words, they specify the conditions of the production of historical knowledge. This amounts to "doing epistemology." Epistemology is once again demystified since even young children are able to engage in this kind of practice. We recognize that it is not self-evident from this example that doing epistemology is a first step on the road to emancipation. However, what happens when older students enact this practice in the context of their science education activities for a significant amount of time?

There are a good number of documented cases of high school students enacting epistemological practice (Larochelle & Désautels, 1992; Désautels & Roth, 1999; Richmond & Kurth, 1999). Through a diversity of activities (writing essays or journals, participating in simulations or small group discussions, participating in research projects, etc.) they had the opportunity to reflect on the "ins" and "outs" of the production of scientific knowledge. In other words, they could step aside and shift their focus so as to transform scientific practice into an object of inquiry. Let us examine how this type of inquiry leads them to question the epistemological status usually attributed to scientists and scientific knowledge, thus making their first steps in an emancipatory intellectual journey.

Although engaged in quite different pedagogical experiences, there is a family resemblance between the discourses of two groups of students, as they become aware that school science has barely anything to do with science in action. One group participated in a seven-week summer camp and collaborated with senior scientists in research activities (Richmond & Kurth, 1999). The other group, almost ten years earlier, participated in a simulation of the production of scientific knowledge (Larochelle & Désautels, 1992) during a one-semester course. The following excerpts indicate quite clearly that following their experience these adolescents (16–18 years old), in both cases, question the conventional ideology of school science through a comparison with the sciences in the making. They no longer believe that scientists are suddenly inspired, that one stumbles on discoveries, or that all experiments work at once. They use different discursive resources to indicate that, trial and error, time and hard work are necessary ingredients in the production of scientific knowledge.

> Science isn't as ideal as it seems in—I guess researchers went through a lot of pain to get those facts that are in our textbooks. And it seems like an inspiration just popped into their heads and the next day they were doing an experiment, and the experiment turned out perfectly the following day, and they got it published. (Daniel, quoted in Richmond & Kurth, 1999, p. 686)
>
> At first, I thought it was something like an inspiration from heaven. I rapidly changed this simplistic view of the process of production. [. . .] To me, scientists were geniuses, two to three times more intelligent than we

were. My idea was that they woke up one morning and said to themselves, "Today, I have this problem to solve." They would then sit in
front of a piece of paper and their intelligence would function by itself.
They then produced scientific knowledge. But, from my own experiences,
I realized that it was not that way at all. You have to work, go on trials
and errors; it is by working really hard that you can arrive at something.
(S-9, quoted in Larochelle & Désautels 1992, p. 235)

These students are now able to make a distinction between
school science and the sciences in action. But they are also quite
capable of conducting a fine analysis of their school experience
and, for some of them, of questioning its formative value. They
notice that school science is dogmatic and does not leave much
room for open inquiry. More so, those for whom this type of reflection was at the center of the pedagogical activity (in the simulation) become quite critical of their school science education as
illustrated by the following excerpts.

Doing freelance research is far more difficult than in school. Because like
I said, when you're in school, the teacher gives you a lab manual, and
says, "Do what it says. Write the data you collect, turn it in." And you
get a grade for it. And there might be a little explanation as to what happened, and everything, but it's something that someone has already done,
so the answer is already there. Right, wrong, or otherwise, you're gonna
know what's gonna happen. So that's one thing I have definitely learned
about. (Chereese, quoted in Richmond & Kurth, 1999, p. 685)
    Personally, I did not think that there could be a difference between
the realization of a research and its public presentation. I thought that
what was presented to us [in school] was truly the research itself. My
perception is certainly due to the fact that science courses today are pro-
grammed and designed in advance. That is why when we carry out an
experiment in any of the fields of science, we are always faced with a
recipe or an outline that we have to follow step-by-step in order to come
up with the expected results. This practice makes us perfect little robots
that are rewarded with marks on their transcripts. I do not think that we
can call that science. (S-21, quoted in Larochelle & Désautels, 1992, p.
223)

As we can see, doing epistemology makes these students sensitive to the artificial and artefactual character of what is being
taught to them and question this type of education. For instance,
it is apparent that S-21 becomes conscious that his or her prior
experience in science education had fostered his construction of an

illusory representation of sciences-in-the-making. He or she is also able to point to the fact that school science is a kind of simulacrum, which conditions students to act as "robots" valuing marks, which is an antithesis of reflexivity and inquisitiveness. From then on, he or she cannot only transform his or her representation of the sciences and "rapport au savoir," but can also call into question the legitimacy of an institution fostering this kind of overdetermination of students.

The students in this particular experiment tackled an impressive number of topics touching upon the logical and social conditions of the production of knowledge. A good number of them could foresee that scientific knowledge is the result of social production and thus relative to its conditions of production. This did not lead them to adopt a solipsistic position concerning scientific knowledge. We could see, however, how they began realizing that much was socially at stake in that enterprise.

> Up until this day, for me, everything within science was certain, what was publicly declared as a law of physics, for example, was infallible. I believe that it is a myth that is widely spread in school about the so-called exact sciences and that I took this myth for reality. Now I can assert that I trust "official" scientific knowledge, but I will never think again that there are absolute truths since I have understood a little better the path one has to follow to be able to produce a piece of scientific knowledge. I had not yet realized that there is no such thing as a great book of nature, which gives you an assurance of the value of your hypothesis. (S-12, Larochelle & Désautels, 1992, p. 218)

Here again a student is able to put into question the traditional ideology of school science, the myth of infallible knowledge, but without throwing away the baby with the bath water. She says that she trusts official knowledge while developing a reservation about the possibility of absolute truth. This can be interpreted as the beginning of a significant transformation in her relationship to knowledge and, consequently, to those who own that knowledge in our society (i.e., scientific experts). Because she also reconceptualized the notion of objectivity as a social process, it is plausible that she is developing the potential for action. This allows her to question the word of experts and participate on equal grounds in whatever sociotechnical debate in which she might

eventually be involved. But this is a story that has not been documented, and why we think that these students are making their
first steps in a journey leading to empowerment.

Intermezzo

Stephen: I think it is clear that in doing epistemology, as participation in a language game, children of all ages forge instruments, potentials for action that can be used to call into question taken-for-
granted ideas about the production of scientific knowledge. But it
seems to me that in the illustrations we have provided, matters of
epistemology tend to overshadow the ethical and esthetical dimensions inherent in the constitution of life forms.

Jacques: You are partly right, but we could see how some of
them started to challenge the quality of their educational experiences, and this is no small accomplishment. However, this impression you have that epistemological matters were at the forefront is
probably the result of our own analytical biases or blind spots.
Let us not forget that in our own education we have been induced
to separate epistemology from ethics as though those categories
were not themselves social constructions. For instance, the fact-
value distinction is so entrenched into our intellectual habitus that
we can hardly imagine that ethics is embedded in knowledge production up-stream as well as down-stream. This is why the ethical
problems as they are related to the sciences are most of the time
cast in terms of the good or bad applications of knowledge.

Stephen: When you say that ethics is embedded in knowledge
production up-stream, I suppose you are referring to the metaphors that guide the research such as the survival of the fittest, the
mechanical worldview or, these days, the idea of a deterministic
genetic program.

Jacques: This is one way to frame the problem. But we must
not forget either that the establishment of scientific disciplines as
dominant regimes of discourse necessarily excludes other forms of
knowledge which are then considered, often in a paternalistic
fashion, as traditional knowledge. At the same time we can see

how the people producing those knowledge are somehow socially disqualified in the name of a set of cognitive values.

Jim: We therefore have to clear up these matters and one good way to do it is to refer back to wine production.

Jacques: Wine production!

Jim: Remember when we stated that the reification of knowledge was like confusing wine in bottles with grapes in the vineyard. The same kind of error is made when we separate facts from values because we then take facts as givens instead of products. In doing so we ignore the whole manufacturing process and render invisible whomever is driving the process, as though facts came from nowhere! We are left to postulate that scientists have developed some kind of magical power (or powder!) allowing them to manifest godlike qualities. Latour (1999b) says, we consider them as the only human beings who can go in and out of Plato's cavern, enabling them to tell the less fortunate, enslaved in the cavern, what the world out there is all about. But if we reintegrate the production of facts in their sociocultural context, at the very heart of the scientific controversies constituting their "native environment" (Stengers, 1997), we can understand that the fact-value dissociation is an artifact in a particular epistemology.

Jacques: I understand what you're saying, Jim, but are we not going to hear the same old story about the matters of fact and the existence of a reality out there made of chairs and tables which have no relation to ethics and values?

Stephen: You are quite right in saying that the realists are going to use the "furniture argument" (Ashmore, Edwards, & Potter, 1994) to counter the socioconstructivist position. They do so as though hitting a table with a fist is not in itself an argument in a rhetorical game but the ultimate gesture pointing at a reality per se. But this argument has been undercut so may times that I wonder if it is worth losing much time and energy to reactivate the outdated realist-constructivist controversy. The work done in the social studies of the sciences over the past 25 years has given enough credibility to the idea that it is viable and fruitful to consider the sciences as embodied, symbolic, material and discursive social practices, that we should not bother with those positivistic reminiscences. I much prefer von Foerster's position (1992). He claims that if we are observers among others in the world, instead of lonely ones, we become responsible for the type of knowledge we produce. Ethics then becomes co-constitutive of knowledge production.

Jim: Even though I tend to agree with a lot of what you just said, we must not forget that positivism nonetheless remains culturally dominant. It is as though the philosophical lessons of Quine (1953) and Duhem (1906/1974) that I summed up for educational researchers (Garrison, 1986) and which delivered a fatal blow to the positivist thesis had not migrated outside the confines of the community of philosophers. Both of them showed that no scientific proposition is testable in isolation independently of the theoretical structure of a disciplinary field in the sciences. I also showed that both fact-theory and fact-value dualism cannot stand scrutiny. Therefore, norm-neutral science and technology is a positivist inherited cultural credo followed almost religiously by bureaucrats and technocrats who construct the state-approved science curriculum, prepare norm-referenced tests, and administrate our schools. No wonder that conflating cultural norms and kinds of expertise will seem heretical to them! We must not forget that they are the power brokers in the system.

Stephen: We could go on with this discussion, but I would like to remind you both that we have to go back to the classroom and substantiate our claims about the enaction of epistemology as social action. But before we describe the next pedagogical episode I would like to bring the following to your attention. We need to consider that helping students to challenge the modern settlement (Latour, 1999a) that separates scientific knowledge from ethical and aesthetic knowledge and thus, the established social hierarchy of knowledge and the powers that come with it, cannot constitute anything but a long-term enterprise. We should therefore be modest and present the episodes as indicative of what can already be done here and now in a certain direction.

Norms in Science and Social Studies Education

The idea that science education is imbued with moral and political norms is best illustrated by the research done in the context of other, non-Eurocentric, cultures. As many science education researchers have concluded (e.g., Baker & Taylor, 1995; Jegede, 1997), the standard science education curriculum imported and enacted in many countries resulted in a rather unfair clash of worldviews which clearly has ethical and political dimensions. In order to perform in the science education system, students generally have to deny the value of their own cultural ways of knowing and making sense of the world. They therefore run "the risk of being viewed as outsiders by their family and community" (Gallard, Viggiano, Graham, Steward, & Vigliano, 1998). The following research conducted in a community on an island in the South Pacific (Waldrip & Taylor, 1999) serves to disclose the phenomenon at hand.

A group of Melanesian students, in residence at a high school on one of the main islands, was interviewed. Much to the researchers' surprise, most of the fifteen students could not provide the traditional explanations that the village elders had given about natural phenomena—as though the students had forgotten their elders' narratives. But as the researchers acknowledge, there might be a number of reasons to explain this situation including the foreign culture of the interviewer and taboos associated with the revelation of traditional knowledge. However, in the course of the interviews, it became apparent that the students seemed to think that the indigenous explanations were ridiculous and characteristic of their parents' ways of knowing rather than their own. More so, "All students except two seemed to feel that the village stories were foolish, and, when pressed for an explanation of natural phenomena, tended to laugh and claim not to know them" (p. 293). Whether they had really forgotten these explanations or not is an unresolved matter for this study. However the fact that they felt embarrassed when speaking of this subject gives an indication of the impact of the symbolic violence (Bourdieu, 1994) exerted by the school on the representation these students have constructed of themselves and their culture. Having unconsciously accepted school science as a criterion to evaluate their own cultural knowledge, they reproduced the relation of domination between worldviews. In other words, the ethical principle of epistemological symmetry was not applied in their science education experience; their ways of knowing have been ignored and despised by the school.

Their lack of indigenous knowledge and their negative appreciation of their own original cultural milieu were also noticed by some of the elders in the community as expressed by two informants.

> Laki: [The younger generation] did not know the old ways. They
>   see them as foolishness. They think they know better.
> Lapun: The young people think that the old ways are rubbish.
>   (Waldrip & Taylor, 1999, p. 295)

Even though one can interpret the elders' comments partly as a manifestation of the usual generation gap in any society, it seems obvious that the net effect of the socialization of these students into the western scientific worldview constitutes a form of cultural alienation. Not only have they been severed physically from their village, they have also been severed from their culture. This situation becomes even more tragic when one realizes most of them will not go on studying after high school but will return to live in their village; this is a rather difficult and troubling experience for them and the community. One of the students said, "Some students, when they go back to the village, they just do sorts of things that village people don't like" (p. 298). More so, as it is generally acknowledged by many informants, school knowledge is more or less useless for living properly in the village. Another student in the group vividly expressed the value conflicts inherent in the situation.

> Because school only helps in the village if you have money. If you don't
> have money, traditional skills and knowledge are far more important.
> Because you can do things, all the resources are there. If you don't know
> how to handle them, and, say, build [a] house with local bush materials
> and all this, it would be quite hard for you to survive in the village. (p.
> 299)

Ethical and power dimensions associated with science teaching are manifest in situations like the one described above. They also pervade our schools but are invisible to us as the everyday categories (Larochelle, in this book) we use to make judgements and take action. The subgroups (cultural, racial, economic, etc.) that perform our societies also construct particular sets of values, beliefs and languages (Hodson, 1999; Ogbu, 1999). This influences their capacity to cope with a school curriculum that reflects the dominant power/knowledge relations. In this sense, schooling in its present form is, for a very large part of the population, an invitation to assimilate a regime of discourse, which conceals itself as an expression of one particular cultural point of view. In this context, science education produces a real hecatomb in the ranks of the students. Most of the victims will not recover from the symbolic wounds that were inflicted on them (Trabal, 1997). We detect something else here as well. School science—as with school knowledge more generally—has no intrinsic value for students. It is only valuable if they can trade it in for "money." In other words, it only has exchange value. They can exchange it for the power and privilege available in the dominant culture, but only if they are docile. Clearly, there are ethical and political values and issues involved here, not just cognitive questions.

In order to take into account the ethical and political stakes involved in such situations, the border crossing metaphor developed by Giroux (1992) offers a promising avenue. The no-assimilation principle embedded in the metaphor, as elaborated by Aikenhead (1996), resonates this recent sensitivity of science educators towards delicate socioethical problems.

> Border crossings may be facilitated in classrooms by studying the subcultures of students' life-worlds and by contrasting them with a critical analysis of the subcultures of science (its norms, values, beliefs, expectations, and conventional actions), consciously moving back and forth between life-worlds and the science-world, switching language conventions explicitly, switching conceptualizations explicitly, switching epistemologies explicitly. (p. 41)

This translation of the no-assimilation principle, which is equivalent to the ethical principle of epistemological symmetry, has lately inspired the work of a number of scholars (Eisenhart,

Finkel & Marion, 1996). These scholars, in their own way, have tried to bring significant changes in the pedagogical practices toward a more critical and democratic form of science education. The following pedagogical illustration, drawn from the social studies education domain, is indicative of this ideological orientation.

Teaching to Develop an Epistemological Stance

It should not come as a surprise by now that eleven- and twelve-year-old middle-school students can "do epistemology." In the present case, students actually studied various aspects of society such that it led them to construct epistemological potentials of action. This transformed their posture towards personal involvement in the use of knowledge. Courses were framed by first posing problems such as, "How do we know something is true?" Initially surprised by radically different teacher expectations, students engaged in discussions about the role of authority, common sense, metaphysics, tradition, intuition, and experience as our basis for knowing anything. Subsequent course activities examined thematic questions about ways of knowing in politics, economics, and society. The subtle pedagogical shift emphasized some of the epistemological issues implicit in typical school subject matter knowledge. This had a significant impact on the students' relationship to knowing and initiated their emancipation from a predetermined epistemology, which called for their passive acceptance of whatever they were told by school and textual authorities.

Having students involved in doing epistemology helps them develop as active members of a participatory democracy. Democratic citizenship education, as it is traditionally conducted by schooling in general and social studies in particular, disempowers students through its emphasis on learning about the structures and rituals of existing institutions and roles. It should, in contrast, develop an understanding of how knowledge serves as a currency in negotiating personal and social power dynamics. For example, within democracies, competing knowledge claims vie to become authoritative policy. Which claims carry the greatest weight in a policy-making process (whether in a family, group of friends, community, school, or a nation) depends on many factors, including the social status of those who are making claims. In contemporary industrialized society, scientific claims (or statements positioned as scientifically based) carry greater weight of expertise. The ability to frame and pose questions about the validity, evidence, assumptions and implications of expert knowledge is an epistemological tool that affords opportunities for non-experts to participate in policy decisions that would otherwise be left to fewer people, ultimately balancing the power of knowledge in society.

One activity that initiated students into thinking epistemologically directed students to discuss the types of evidence needed to determine each statement on an "epistemology sheet" as "true" or "false." Additionally they were to identify at least one potential weakness of their suggested evidence. The list included statements such as "Women shop for groceries more often than men," "The United States became a country in 1776," "Angels are smaller than most human beings," "Rich people have more free time than poor people," and "The United States has always been friendly to Canada." As students discussed in small groups, issues about validity and reliability arose over what could "count" as evidence. In subsequent class discussions, students categorized different types of evidence as "proof by authority," "tradition," "intuition," "superstition," and "science" (observation).[1] It is important to note that students constructed these categories. Typical responses included, "Many things can't be proven by other people if they can't be seen," "A lot of knowledge is information from someone else, not from what you see," and "People don't always agree about the same evidence." Some students recognized that evidence based on "seeing" had a better chance of being agreed to by more people. We used this as a chance to explore differences in

---

1. Roth and Lucas (1997) showed that these and other "categories" can be viewed as discursive repertoires. When asked to justify ontological, sociological, or epistemological claims, students draw on these for-the-moment unquestionable repertoires to support the more tenuous claims.

using "private knowledge" such as intuition or superstition, compared with "public knowledge" (e.g., science) in a democracy. Subsequent interactions revealed that student postures toward knowledge seemed to change.

Doing epistemology is more than demonstrating a set of technical behaviors in isolation from a broader set of ideas, beliefs and competencies about knowledge. It involves a comprehensive change in how students come to see and use knowledge, a planned and deliberate development of ideas and practices about knowledge production, and transformation in students' postures toward what counts as worthwhile intellectual work inside and outside school. Giving students "permission" initially to challenge official school/social knowledge has to be accompanied by structured pedagogical experiences where students gain the confidence to assume permission. The following case is selected from a teaching project over a three-year period that involved students in questioning and challenging social knowledge.

The curricular goal of studying the "creation myth" of Native Americans indigenous to the students' geographic area provided a classroom opportunity for students to become aware of the dominance of their own cultural ontologies and to be introduced to the idea of epistemology. Initially regarding the Iroquois explanation of the creation of the world as strange and inferior to their own scientifically flavored narrative, students were provided creation explanations from twelve different cultures to examine, including the traditional Judeo-Christian version and the more recent version of contemporary science. In planned, cooperative group activities, students were guided to note differences and similarities among creation myths and to hypothesize about the relationships between specific cultural conducts and characteristics in the myths. During their small group discussions and in subsequent whole class discussions, students generally concluded that most people use their experiences and what is familiar to them to explain why things are the way they are.

As part of the activity, students were encouraged to consider the value-laden basis of various ways of knowing. Students were asked first to describe each creation account, then to identify on what basis the society thinks they "know" the account true, and finally to discuss why the society might believe their thinking about what counts as truth is right. As one might recognize, the third question was the most difficult question to consider, and one for which many students needed to be prodded, but it inevitably lead students to examine the faith involved in contemporary accounts.

A set of interactions at one table exemplifies the discussions about the truthfulness of scientific and religious accounts of creation in comparison to that of the creation myths.

Matt: We know about this because of scientists and all their experiments!

Art: But if you think about it, how can they really do experiments on something like this?

Matt: I don't mean like with test tubes or anything like that. I saw on television that scientists can tell by looking how fast the planets and stars and everything else is moving.

Art: Yeah . . . but maybe everything is just racing around up there . . . no reason, you know. If you don't hear it or see it yourself, if no one heard or saw anything, how could they really know something like that happened?

Teacher: Excuse me, I overheard your conversation and wondered about our own bible explanation? How does it fit with all these other stories?

Tina: I don't see how the Bible is really much different. I mean, some of the stuff in the Bible sounds like the other stories we talked about, you know, riding on a turtle's back [Iroquois] and things like that- it's not stuff that would probably really happen in real life. But the main part of the story, where the world is created, maybe that's real. Maybe God made the big explosion, then he probably came around afterwards to clean up things!

Art: Oh yeah? He hasn't done so hot lately. You ought to see the buildings on my road!

Teacher: You're raising some good points. But do you think our views [the bible or science] are more truthful than the creation myths we've looked at?

Tina: Do you mean by truthful if we know for sure . . . ?

Matt: In this class, I don't think I know anything for sure anymore!

Teacher: I'm just asking if these stories all make the same kind of sense to you? Do they look equally truthful?

Art: I'm kind of teasing Matt about scientists because I know he's crazy about science! But that [the Big Bang] makes the most sense to me too. It's based on things scientists look at, not

> imagined, but since we've been talking about it in the last few days. I was thinking maybe I like our view because it's our view, but us students can't prove it, we're not scientists. We have to believe them, so I can see how other stories make sense to other people.

Matt: Right! So even though it's science, we still have to believe in some things without knowing for sure!

An important turn of mind should be noted here. Myth, used in this way, went from being viewed by students as simply a fantastic story about something untrue to a plausible explanation based on the truths and beliefs available within one's personal and cultural experiences to make sense of the world. Additionally, some of the students (Art and Matt in particular) were beginning to consider that science includes human moral and aesthetic values, not just epistemic values.

A subsequent activity involved students in composing their own creation myths, drawing upon their own experiences and their interpretation of other accounts. In this activity, the process of thinking and writing was the dominant content goal. The use of a four-step writing process to construct their creation myths was an epistemological activity meant to initiate a further shift in valuing their own knowledge. Advising students to "be messy, rough, cheat and copy" elicited great interest, and helped them break the idea that thinking, writing, and knowledge production "just happen for smart people." For example, being "messy" freed them up to list all the ideas they could think of in any order, while being "rough" signaled to them the relationships they begin to draw among ideas are tentative and mutable. "Cheating" ironically plays on the idea of having someone else look at your work and editing, which, of course, is an example of the social construction of knowledge. And "copy" conveys to them that products of knowledge construction often conceal how other people's ideas have been used. When asked to comment about what they learned about the process of thinking and writing from the above activity, students offered various comments. These included, "No idea is the only one that is right or correct," "Thinking and learning is a cooperative activity," "Coming up with ideas takes a bunch of steps," and "There are many ways to say the same thing."

The creation myth activities provided a means for viewing science and scientific theories as contemporary mythical explanations about social phenomena, i.e., explanations involving empirical information as well as untested—or untestable—beliefs reflecting cultural-historical values. The purpose was not to diminish students' faith in science or scientific theories but rather to help them begin to develop an understanding that all knowledge, including science, is within a cultural and historical context.

Intermezzo

Jacques: It is apparent that in the process of writing their own "creation versions" to explain the world, the children were involved in more than an orientation to epistemology as "truth-seeking." I mean, the students did not conclude from experiencing this activity that they had a "correct" explanation for how the world came to be! Students began to reflect on the process of knowledge production. They realized there is more than one way to make a bottle of wine.

Stephen: Their involvement offers students an important lesson concerning the role of their own experiences, personal and otherwise, in knowledge making. Doing epistemology involves far more than enacting one's logical capacities. Having personal experiences in knowledge making might also enact and enliven one's aesthetic proclivities. In turn, and with reflection on the personal satisfaction of these experiences, students might develop a different posture towards what Apple (1993) calls the "official knowledge" handed to them in schools.

Jim: I agree. But remember that we agreed to proceed cautiously when discussing our tentatively constructed distinction of ethics, epistemologies, and aesthetics!

Stephen: Yes, we did, and I assume that is why I used the word "posture" instead of "attitude" to discuss the potential effect of our aesthetics perspective?

Jacques: Right! The use of psychological terms such as attitude seems almost natural in a discussion about educational concerns. Given its history over the past century, psychology has contaminated much of our thinking about education. Popkewitz (1998) puts it this way, "The administration of reflection replaced revelation in finding human progress" (p. 89). In other words, psychology has made it possible to secularize the religious motif of the governing of the soul through the naturalization and reification of categories such as attitude, learning, self-actualization, intelligence and so forth.

Stephen: Perhaps so, but for our purposes in this discussion we can agree that "posture" represents something more comprehensive about knowledge than "attitude"? I mean for me, attitude sounds very passive, as if consisting of innate attributes of the knowing subject. On the other hand, posture suggests one is active with knowledge—that somehow emotions, passions, and needs are equally important in the act of knowing. I think that aesthetic experiences are at the root of changing one's rapport au savoir.

Jacques: At the root of changing one's relationship to knowledge? Is it not a consequence? This sounds like a very large claim, if I understand you correctly, and quite the opposite of what we traditionally claim. By implication you are suggesting we have previously ignored what might be one of the most important considerations in our work as educators! It surely has implications for our interest in furthering democratic and scientific participation!

Jim: Well, it is a big claim . . . but an exciting one! Consider how education, like psychology, has been "de-aestheticized" by restricting the experiences students have with knowledge and knowledge making. Providing only the products of knowledge for students prevents them from being involved in the process of making the products. This is the most likely place for creating . . . for stimulating curiosities, for satisfying their need to unify experiences, and for seeking more knowledge to use as an instrument in satisfying even more aesthetic cravings!

Jacques: Can't we see there the result of the cultural hegemony of the Enlightenment's supposedly self-sufficient reason in our educational projects?

Stephen: But you can see why it is such a difficult pedagogical tradition to break when you consider how it emanates from a culturally deep-seated hierarchy of knowledge that even dates to the ancient Greeks! In their view, only contemplative knowledge reveals nature as it really is. Artistic knowledge was considered a lower form because its production is based on an imperfect and subjective world.

Jim: This cultural perspective illuminates a lot about the status of contemporary education. You can see how the legacy of higher and lower knowledge is culturally consistent with our tacit beliefs and practices as teachers. We might believe that we want to teach students how to think and value the process of reflection and really be involved, heart and soul, in understanding. But our tacit belief—the one that is culturally embedded and overrides all other beliefs—is that there is essential, contemplative knowledge for students to "know." What we mean by "knowing" is tantamount to a commodity, not a relationship!

Jacques: This may be the legacy of centuries of catechism! Perhaps for similar reasons standardized examinations and national standards are co-incidentally resisted and embraced by educators . . . resisted because they restrict the creative prowess of teachers and students but embraced as modern manifestations of contemplative knowledge?

Jim: Pardon my slight change of tone—or perhaps I mean posture—but we can see now some of the barriers we are up against. So when we look at the possibilities for science education that are democratic and liberating, it seems we agree on the following. A more substantive exploration, or perhaps exploitation, of aesthetics might offer an antidote to some of the damage that has been inflicted on education through our traditional categories and distinctions.

Stephen: I can think of an example. I was just re-reading C. Wright Mill's work on the sociological imagination, the part where he refers to C. P. Snow's two cultures thesis. When viewed aesthetically, as human meaning-making activities, the traditional distinction between "science" and "art" disappears. . . .

Jim: Which reminds me of Jacob Bronowski's point in *The Ascent of Man* [sic]. What a riveting experience!

Stephen: That was Bronowski's theme in all his works. Remember how the classical distinctions between "humanities" and "sciences" collapse in his treatment of "Science and Human Values"? Using aesthetics as his entry, Bronowski exposes science, or more precisely, science-as-a-way-of-knowing, as perhaps the most humanistic of the humanities!

Jim: Well, Dewey reminded us of this earlier when he says that the process involved in the creative work of an artist is found in all intelligent, creative human activity. Science is an art whose artifact is "truth"—what he prefers to call "warranted assertability" to underscore that knowledge is contingent. Pragmatists and critical constructivists embrace our imperfect, subjective worlds as the best ones, perhaps the only ones, available for us to make meaningful lives.

Jacques: So the fact-value distinction that usually accompanies the common separation of the arts and sciences turns out to be a misguided simplicity! Our aesthetic lenses help us see their commonality. But what does it say to us about science education in a democracy . . . about critical scientific literacy? What about those of us who are not artists or scientists? How do we have access to these aesthetics? Our discussion still seems to keep science distant from the masses of students.

Jim: The notion that only artists can have aesthetic experiences is another ill-fated notion related to our Greek legacy! Remember, *techne* is the knowledge of how *poiesis* or creating, is making or calling into existence. Being an artist or a scientist is an occupation, but intelligent creative human activity is commonplace and can occur in anyone's life. So in that very sense, all students encounter aesthetic experiences in their schooling. Perhaps the issue is to what degree are these aesthetic experiences validated, supported, and used in their education?

Stephen: Yes, that's part of it. Also, the meanings we share in our common exploration of science and epistemology concern our common desires for a more symmetrical balance of power in society when it comes to issues involving decisions of science and technology. Our discussions have always been cemented by our common anti-authoritarian posture, a word I selected to convey the inseparability of our knowledge, passion, and ethics. We see this as crucial for the successful development of democracy, a goal towards which we put logic and science to work not a goal attained through logic or science.

Aesthetics in Science and Science Education

The lost of unity of experience alluded to in the dialogue might just be an illusion resulting from the domination of the rationalistic and positivistic discourse as a referent for describing the production of scientific knowledge. But, as the historian James W. McAllister (1996) has pointed out, aesthetics has always played an important role in the sciences whether we consider the individual or the collectivity. For example, it took Kepler a good number of years before he could admit to himself that the planetary orbits were not circular, because the circle was held to be a perfect "divine" figure. More so, scientists at each historical period collectively adopt aesthetic canons for the production of theories. Thus, while a theory was considered more beautiful the more it was abstract during the 18th century, theories were only considered when they could be visualized and transformed into mechanical models (McAllister, 1996). In a way, even if one does not agree with McAllister's general explanation of the role played by aesthetics in the production of scientific knowledge, he clearly showed that scientists delve in experience just like other human beings. This becomes even more obvious if we follow the intellectual path opened up by Serres and Farouki (1999) in a wonderfully illustrated book that, as they suggest, can be read as an art book.

Serres and Farouki claim that aesthetics is again in the foreground, because more and more scientific knowledge is produced through simulations. Scientists bring forth realities by making choices among many possibilities. In the process, they construct beautiful, detailed and virtual landscapes called molecules, proteins, bacteria, DNA, galaxies. At the same time, they cannot but claim responsibility for what they produce. Serres and Farouki draw a parallel between the production of the artist and that of the scientist. Both follow similar courses of action and have the "same obligations to look at details, similar confrontations with unforeseeable circumstances, common recourses to imagination, same location in a world of possibilities, similar range of choices among virtual worlds, comparable production of the real and, soon, same responsibility" (p. xli). In such a context, the referent is no longer an objective world "out there" beyond the appearances, but the world "in here" as shown in all its details and particularities on the photographs or on the screen. Therefore, the experience of scientists is simultaneously and at once epistemological, ethical, and aesthetic. More so, in a world now conceived as indeterminate, in which the initial conditions or circumstances cannot be judged identical, one cannot predict the outcomes. One has to become provident and follow the principle of precaution. But how are those ideas received in science education?

Professional documents show quite clearly that, at this time, aesthetics has little currency in science education. For instance neither of the most recent handbooks on research in science education (Gabel, 1994; Fraser & Tobin, 1998) has an entry in their index related to aesthetics. In fact, most science education research ignores the aesthetic dimensions of the experience students live through as they try to make sense of their participation in science education. The main focus is put on epistemological or cognitive aspects of their activity. There are probably many good reasons to explain such a situation, but we have not the space to examine them more closely. We concentrate instead on illustrating how aesthetics come into play in the lived experiences of students in science education.

In a study of students' and their teachers' worldviews, Cobern, Gibson, and Underwood (1995) chose the topic of nature to structure interview protocols. Ann, one of the students in Mr. Hess' class, provides a telling example of how aesthetics mediates her experience. Indeed, as one can notice in her discourse, beauty and purity are closely associated to her apprehension of what she calls nature.

> To me nature is beautiful and pure because it is God's creation. Nature provides both aesthetic and emotional pleasure and I need it for self-renewal. . . . When I'm out in nature, I feel calm and peaceful. It is a spiritual feeling and it helps me understand myself. . . . The pleasure I get from nature is enhanced by the mysteries I see in it. (Cobern et al., 1995, p. 24)

If Ann's discourse is compared to her teacher's, it becomes understandable why she thinks the content of her science class has nothing in common with what she calls nature. Here is how Mr. Hess talks about the same subject.

Nature is orderly and understandable ... [and] the planets and the stars are governed by physical forces and any deviations are simply because we have not yet discovered the other part of nature's orderliness.... Scientific or reductionist thinking is very powerful. I feel that, once we know enough about the minutia of the world, breaking it down by using scientific method (with scientists tearing it apart and analyzing the parts of nature and seeing how they interact), we will be able to predict just about anything about nature. (Cobern et al., 1995, p. 18)

What we would like to stress is that these discourses belong to two quite different aesthetic orientations. The first one has artistic and spiritual overtones while the second has a rationalistic and even scientistic flavor. In this particular case, the incompatibility between the two orientations did not prevent the student from doing well in science, but such is not the case with other students who develop a real aversion towards science education, especially when the cultural stakes are high (Cobern & Aikenhead, 1998). In fact, the incompatibility of worldviews and their embedded aesthetic orientations is one of the reasons why students shy away from science and have difficulties in making sense of their science education experiences. Therefore, the principle of epistemological symmetry has some limitations, if we are to take into account the whole cultural experience of the students and if we want science education to become a democratic social action.

Epilogue

In this chapter we uncover some of the conditions under which students can transform their relationship to knowledge to emancipate themselves from their own biographies and become actors in a more democratic society. We described pedagogical situations that illustrate the feasibility of an education oriented to those principles. In other words, we believe that it is possible to enact a form of critical pedagogy in science.

Good teachers want to empower students. Indeed, empowerment is the purpose of enacting epistemological practice as subversive social action. Our goal is to seize school science from the exclusive position of scientific experts and school bureaucrats and technocrats. At the same time, we appreciate the irony of saying

to anyone: I am going to empower you. In its arrogance and conde-
scension, it suggests "the other" is powerless before us. Wherever
there is power, there is paradox, irony, and reversal. Emancipatory teachers must be eternally vigilant least they become the
problem rather than the solution. Nonetheless, we believe the enactment of epistemological practice is a form of social action that
serves to subvert the social hierarchies that define not only modern, but classical culture. We believe allowing students to bring
their ways of knowing, values, and beliefs into a living dialogue
within the classroom and community is the best road to scientific
and technological literacy. Instead of liberal education, we seek a
liberating education.

Foucault did not think of power as a mystical substance. Instead, Foucault thought of power as a matter of relationships,
functional relationships. We live in holistic webs of power just as
we live in holistic webs of belief. It is not a matter of escaping
from all "power"; that is impossible. We can, however, reconfigure
power in ways that make it more functional and less oppressive.

References

Aikenhead, G. S. (1996). Science education: Border crossing into
    the subculture of science. *Studies in Science Education, 27,* 1–52.
Apple, M. (1993). *Official knowledge: Democratic education in a con-
    servative age.* New York: Routledge.
Ashmore, M., Edwards D., & Potter, J. (1994). The bottom line:
    The rhetoric of reality demonstrations. *Configurations, 1,* 1–14.
Baker, D., & Taylor, P. (1995). The effect of culture on the learning
    of science in non-western countries: The results of an inte-
    grated research review. *International Journal of Science Educa-
    tion, 17,* 695–704.
Barnes, B. (1990). Sociological theories of scientific knowledge. In
    R. C. Olby, G. N. Cantor, J. R. R. Christie, & M. J. S. Hodge
    (Eds.), *Companion to the history of modern science* (pp. 60–73).
    London: Routledge.
Biagioli, M. (Ed.). (1993). *The science studies reader.* London: Rout-
    ledge.

Billig, M. (1996). *Arguing and thinking*. Cambridge, England: Cambridge University Press.

Bourdieu, P. (1994). *Raisons pratiques. Sur la théorie de l'action* [Practical reason: On the theory of action]. Paris: Seuil.

Charlot, B., Bautier, E., & Rochex, J.-Y. (1992). *École et savoir dans les banlieues . . . et ailleurs* [School and knowledge in the suburbs . . . and elsewhere]. Paris: Armand Colin.

Cobern, W. W., & Aikenhead, G. S. (1998). Cultural aspects of learning science. In B. J. Fraser & K. Tobin (Eds.), *International handbook of science education* (pp. 39–52). Dordrecht, Netherlands: Kluwer Academic Publishers.

Cobern, W. W., Gibson, A. T., & Underwood, S. A. (1995). *Everyday thoughts about nature: An interpretative study of 16 ninth graders conceptualization of nature*. Paper presented at the annual meeting of the National Educational Research Association, San Francisco, CA.

Delbos, G. (1993). Eux ils croient . . . Nous on sait . . . [Them, they believe . . . Us, we know . . .]. *Ethnologie française, 23*, 367–383.

Désautels, J., Garrison. J., & Fleury, S.C. (1998). Critical-constructivism and the sociopolitical agenda. In M. Larochelle, N. Bednarz, & J. Garrison (Eds.), *Constructivism and education* (pp. 235–270). New York: Cambridge University Press.

Désautels, J., & Roth, W.-M. (1999). Demystifying epistemological practice. *Cybernetics & Human Knowing, 6*(1), 33–45.

Duhem, P. M. (1906/1974). *The aim and structure of theory* (P. P. Wiener, trans). New York: Athenaeum.

Eisenhart, M., Finkel, E., & Marion, S. F. (1996). Creating the conditions for scientific literacy: A re-examination. *American Educational Research Journal, 33*, 261–295.

Fasulo, A., Girardet, H., & Pontecorvo, C. (1998). Historical practices in school through photographical reconstruction. *Mind, Culture, and Activity, 5*, 253–271.

Foucault, M. (1971). *L'ordre du discours*. Paris: Gallimard.

Fraser, B., & Tobin, K. (Eds). (1998). *International handbook of science education*. Dordrecht, Netherlands: Kluwer Academic Publishers.

Gabel, D. L. (Ed.). (1998). *Handbook of research on science teaching and learning*. New York: Macmillan Publishing Company.

Gallard, A., Viggiano, E., Graham, S., Steward, G., & Vigliano, M. (1998). The learning of voluntary and involuntary minorities in science classrooms. In B. J. Fraser & K. Tobin (Eds.), *International handbook of science education* (pp. 941–953). Dordrecht, Netherlands: Kluwer Academic Publishers.

Garrison, J. (1986). Some principles of postpositivistic philosophy of science. *Educational Researcher, 15,* 12–18.

Giroux, H. (1992). *Border crossings: Cultural workers and the politics of education.* New York: Routledge.

Hodson, D. (1999). Going beyond cultural pluralism: Science education for political action. *Science Education, 83,* 775–796.

Jegede, O. J. (1997). School science and the development of scientific culture: A review of contemporary science education in Africa. *International Journal of Science Education, 19,* 1–20.

Larochelle, M., & Désautels, J. (1992). *Autour de l'idée de science. Itinéraires cognitifs d'étudiants et d'étudiantes.* [About science: students' cognitive paths]. Québec/Bruxelles: Presses de l'Université Laval/De Boeck-Wesmaël.

Latour, B. (1999a). *Pandora's hope.* Cambridge, MA: Harvard University Press.

Latour, B. (1999b). *Les politiques de la nature. Comment faire entrer les sciences en démocratie* [Politics of nature. How to bring science into democracy]. Paris: La Découverte.

McAllister, J. W. (1996). *Beauty and revolution in science.* Ithaca, NY: Cornell University Press.

Messer-Davidov, E., Shumway, D. R., & Sylvan, D. J. (1993). *Knowledges: Historical studies and critical studies in disciplinarity.* Charlottesville: University Press of Virginia.

Ogbu, J. U. (1999). Beyond language: Ebonics, proper English, and identity in a black-American speech community. *American Educational Research Journal, 36,* 147–184.

Popkewitz, T. S. (1998). Knowledge, power, and curriculum: Revising a TRSE argument. *Theory and Research in Social Education, 26,* 83–101.

Quine, W. V. O. (1953). *From a logical point of view.* Cambridge, MA: Harvard University Press.

Richmond, G., & Kurth, L. A. (1999). Moving from outside to inside: High school students' use of apprenticeships as vehicles for entering the culture and practice of science. *Journal of Research in Science Teaching, 36,* 677–697.

Serres, M., & Farouki, N. (1999). *Paysages des sciences.* Paris: Le Pommier.

Stengers, I. (1997). *Sciences et pouvoirs. Faut-il en avoir peur* [Science and power: Is there cause for fear?]. Brussels: Labor.

Trabal, P. (1997). *La violence de l'enseignement des mathématiques et des sciences* [Mathematics and science teaching as a source of violence]. Paris: L'Harmattan.

von Foerster, H. (1992). Ethics and second-order cybernetics. *Cybernetics & Human Knowing, 1*(1), 9–19.

Waldrip, B. C. & Taylor, P. C. (1999). Permeability of students' worldviews to their school views in a non-western developing country. *Journal of Research in Science Teaching, 36,* 289–303.

From Science to Epistemology and Back

Wolff-Michael Roth & G. Michael Bowen

> And from experiment we can prove that energy and entropy really do exist. Atoms have always been hard to see because they are so small. But with today's major advances in science we are able to see with high power microscopes that they really do exist, and also most of the characteristics that were said to be true, after years of experimenting by many scientists, were indeed true. (Kyle, grade 12 physics)

In this statement, Kyle (pseudonym) made claims about the nature of "energy" and "entropy" on the one hand, and about the existence of "atoms" on the other. He buttressed his claims by drawing on what has been termed an *empiricist* interpretative repertoire (Roth & Lucas, 1997), that is, he supported his claim that energy, entropy, and atoms *really* exist by suggesting that there is experimental evidence that *proves* this existence. In this case, the student regarded the claims themselves as needing support but drew on another aspect of discourse in support of the claim. Those discourse aspects which are taken for granted and are unquestionable at a particular moment, and which therefore can be employed in support of more tentative claims, are generally referred to as *discursive repertoires* (Potter & Wetherell, 1987). Recent research in discursive (social) psychology (e.g., Edwards & Potter, 1992; Harré & Gillett, 1994), social studies of science (Mulkay & Gilbert, 1983; Latour & Woolgar, 1986), and science education (Roth & Alexander, 1997; Roth & Lucas, 1997) suggests that it is more interesting to study the nature of the interpretive repertoires than the content of talk. Discursive repertoires are said to be more interesting because they specify what members of some community regard as their common ground on which they can rest their more tentative claims and points of contention.

Scientists (and science teachers) view (teaching) understanding nature through facts, laws, and theories as their core purpose (e.g., Kuhn, 1970). Epistemologists (e.g., von Glasersfeld, 1989; Fuller, 1988) and sociologists (e.g., Latour, 1993b), on the other hand, are interested in the (social) construction of claims about the nature of this knowledge. Both processes are implicated in the same practices of science, that is, each construction of a scientific fact implies a particular epistemology. Yet despite calls to include epistemology (nature of science) in science lessons (e.g., Larochelle & Désautels, 1992; Lederman, 1992; Roth & McGinn, 1997), very little of such teaching appears to occur in practice. Students are made to learn "facts," "laws," and "theories" but do not have the option of questioning any one of them. In this (learning about) science differs very little from doctrine (Fuller, 1997). Even if students were to question some law or theory based on their own investigations, teachers (and professors) are likely to draw on their (institutionally derived) authority as a repertoire to dismiss such claims. Such incidences occur both at the high school and university levels (e.g., Roth, McRobbie, Lucas, & Boutonné, 1997; Roth & Tobin, 1996). Yet there is evidence that high school students are capable of engaging in epistemological discourse as practice (Désautels & Roth, 1999). Furthermore, such discourse, in the context of science courses, allows students to take a reflective and knowledgeable stance with respect to the nature of knowledge and the role of claims and evidence.

This chapter is part of our overall project to construct an argument for teaching (Western, local, informal, or indigenous) science not just as bodies of knowledge, but as an engagement of students in a dual journey during which they learn to talk science *and* epistemology. In this chapter, we trace out the following itinerary. After reviewing the literature on claims and evidence, we provide exemplary episodes from three quite different settings. These include (a) professional science (an ecological field camp), (b) a grade 8 class studying ecology by designing their own research programs, and (c) a grade 13 class studying epistemology as part of their advanced physics course. Because the same researchers and assistants conducted all three ethnographic studies, (rarely possible) comparisons can be made (are interpretive patterns transportable across contexts?), the ethnographic equivalent of "generalizability" (Guba & Lincoln, 1989). These case studies illustrate similarities between scientists and grade 8 students as to how they constructed claims and evidence, and differences between these two and a group of physics students who also engaged in epistemological discussions about knowledge claims and evidence. Finally, we conclude with a call for an approach to science teaching that integrates the subject matter and epistemology.

Claims and Evidence

In many scholarly circles, it has almost become a truism to state that descriptions of scientific facts cannot be grounded in an appeal to nature. Many philosophers agree that all there is are networks of (discursive, material) practices, which, in their interactions, are stabilizing each other (Quine, 1995; Rorty, 1989; Wittgenstein, 1994). This is, because there can never be raw observation; all observation is "theory-laden" even if this theory is embodied in our daily common sense (Pollner, 1987). It therefore comes as no surprise that social scientists have begun to more closely examine forms of discourse and the (shifting) relations between scientific claims and evidence made by scientists and, in public debates, by politicians, lawyers, and "just plain folks" (e.g., Epstein, 1997). In this research, it is generally assumed that we can never know "knowledge," "beliefs," "interests," "attitudes," and so forth other than through the discourses people engage in. Because it is difficult to ground this discourse by an appeal to nature, it is therefore more interesting to study what are taken to be the common unquestionable (discursive) grounds on which claims are rested, and the discursive devices employed to reconcile incompatible claims.

In a study of microbiologists, the emergence of facts from the uncertainties and contingencies of laboratory work has been described in terms of a process of "hardening" (Latour, 1987). Accordingly, depending on the discursive support available, scientific statements can be framed "softer" or "harder," leading from a "weaker" to a "stronger" rhetoric, that is, the modalities change with the available discursive repertoire. Depending on the presence or absence of modalities, statements are acknowledged along a gradient from hard fact to artefact (e.g., scientific hoax in general, and cold fusion and N-rays in particular [Lewenstein, 1995; Schnabel, 1994; Simon, 1999]).

A seminal study of the language used in an open scientific debate showed that scientists expressed themselves differently about the factual nature of a chemiosmotic phenomenon in formal and informal situations (Gilbert & Mulkay, 1984). Across these situations, scientists' statements were frequently inconsistent. For example, across the two types of situations scientists claimed, on the one hand, that their results supported a particular, controversial model (theory). On the other hand, the same scientists claimed the reverse: the model suggested the results. In another case, the same scientists suggested that science is both rational and irrational. Based on the assumption that people do not attempt to be irrational, Gilbert and Mulkay proposed that we should consider their talk in the form of two repertoires: formal-empiricist and informal-contingent. When their claims from the two repertoires were different, scientists used the "truth-will-out-device" according to which truth (fact) will arise from the contingencies of daily scientific work.

In modified form, these sociological approaches have been successful in understanding the epistemological, sociological, and ontological discourses (and changes therein) of high school physics students (Lucas & Roth, 1996; Roth & Alexander, 1997; Roth, McRobbie, & Lucas, 1998; Roth & Lucas, 1997). In our work, we therefore adopt a pragmatic approach. Accordingly, language is treated neither as if it expressed or represented reality nor that it stands between individual Self (mind) and Other (physical and social world): Language is all there is. Thus, human beings use language to weave and reweave "beliefs," "desires," "attitudes," "Self," and "reality" (Derrida, 1988; Rorty, 1989). In adhering to this maxim in our research, we heed the ethnomethodological advice to "examine the ways that conduct, belief, and judgement are organized, produced and made intelligible in members' own accounts and descriptions" (Jayyusi, 1991, p. 234).

In the following three case studies we exemplify what we have learned from our own research among scientists and in schools about how claims and evidence are constructed. Because we conducted all three ethnographic studies, we are in a singular position to make comparisons across interpretive research and thereby analyze the transportability of qualitative patterns. Because few researchers actually get to enact ethnographic studies among diverse participants, such comparisons are rare, which makes this paper unique in many ways. The comparison between our three exemplary episodes brings out two main features. First, there are similarities between the claim and evidence construction between the scientists and students taught according to the "authentic science" metaphor. Second, there are differences between these two groups and students who also discuss epistemology along with learning science. In the former case, claim and evidence are constructed according to domain-internal rules, whereas epistemologically based discussions also encourage questioning (not only of scientific but all sorts of) claims and evidence from the outside. Our ultimate goal is to suggest that in democratic societies, the evaluation of claims and evidence cannot be left to domain-internal criteria but also requires cross-domain criteria thereby decreasing our dependence on so-called "experts."

Claims and Evidence in Scientific Practice

In this case study, we exemplify what we learned about the construction of claims and evidence in one domain of science.[1] Over a two-year period, we conducted an ethnographic study among ecologists including field observations, formal and informal interviews about their work, and formal interviews about graphs culled

---

1. Because of space limitations, we cannot detail the research methods enacted in each study. Details of the extensive data collection and interpretative procedures are found in Roth and Bowen, 1995 (grade 8 study), Roth and Lucas, 1997 (grade 11 study), Roth and Bowen, 1999b (ecologists).

from an undergraduate course and from their own work (e.g., Roth & Bowen, 1998, in press; Roth, Masciotra, & Bowen, in press). In one case, we followed a scientist (Sam) through two field seasons and conducted extensive interviews before and after fieldwork when she was back in her laboratory. We also attended five local, national, and international conferences where she presented papers based on her work. In this study, we learned that what ecologists reported as scientific facts arose from initially very tentative statements, which, in the course of their inquiries, become increasingly hardened. Furthermore, we could observe the practices surrounding evidence construction that resulted in a shoring up of evidence. As a consequence, other researchers would be less able to attach modalities that diminished or undermined the facts constructed.

Hardening the Evidence

One of our key informants was Sam, a doctoral student who, now in her sixth field season, had extensive research experience. The following transcripts recorded at various stages from hunting for lizards in the bush to presenting the research results of that work to a scientific audience.

> [Out in the bush:] I usually find about five [lizards] a day. I sort of am getting this feeling that they are more active later in the day. They can't tolerate, I think preferred temperature is about 20, mid 20's or maybe high 20's. Probably mid. . . . So in the real heat of the day I don't look for the animals 'cause they're buried down too deep and then I go out again in the 4 to 6 [p.m.] kind of range and lately I've noticed I've had better luck.
>
> [In the laboratory:] I don't know if I will be able to use these speed measures, but I do it anyway. Maybe there is something, maybe not.
>
> [At a conference:] And it turns out the longer the lizards are kept in the lab, the slower they run. . . . And adult males also have relatively longer back legs than adult females. And it turns out that this body length and back leg length are important for predicting how fast it runs.

We can see here that Sam's discourse makes use of modalities that contribute to different levels in the hardness of her statements. In the first statement, the "facts" are very soft; Sam frequently referred to the kind of understandings expressed in these statements as "anecdotal." The second statement shows that she already had framed a variable, sprint speed, which was possibly related to other characteristics of the lizard and an indicator of the natural history of lizards. In the final statement, recorded during a conference presentation, Sam makes a definite statement about the relationship between body and leg lengths and sprint speed. There are no longer qualifiers such as "I think," "Maybe, maybe not" and "probably" that contribute to a softer modality.

Our formal interview study further showed that whereas scientists often disqualify their "anecdotal" knowledge, it was this selfsame knowledge which allowed them to interpret new and unfamiliar graphs. That is, when scientists could not draw on some form of embodied knowledge (relating to either field research practices or mathematical transformation practices), they failed to provide scientifically correct interpretations (Roth & Bowen, 1999a).

During our fieldwork, we had many opportunities to observe that the construction of claim-supporting evidence was not only implicit in scientists' practices but also frequently addressed in explicit ways. Although our field ecologists were not in a situation where they had to defend their claims, they often did not even have tangible claims and were continuously concerned with issues of reliability, repeatability, and consistency. That is, they were engaged in the production of scientific knowledge but had to work around and within the local contingencies of equipment and instruments. In this way, our scientists had to balance local solutions with the anticipated, evaluative conventions of a wider scientific community and audience.

This is especially the case when a scientist has little familiarity and experience with some measurement instrument (e.g., Nutch, 1996). One of the cases we observed comes from our first field season while observing and recording Sam at work. At this time, she was just beginning to use a Munsell chart to determine the color and mottledness of the lizards that she had captured.

Each page of the Munsell chart consists of color plates in the center of which there is a viewing hole through which the object of interest is examined. In this way, each viewing hole of the Munsell chart already provides a context-free reference and abstracts the object of inquiry from the surrounding context. When used for their original intent, classifying soil color in the field, moistening of the soil samples brings about an additional uniformity among different objects.

However, doing a color determination of the lizards in the field was not sufficient for Sam. She did the color determinations in the lab, although from a perspective of adaptation and evolution, the color of a lizard in its natural environment is the crucial aspect. From the beginning of her work with the Munsell chart Sam enacted particular concerns related to the variability of color determination and color variability setting up conditions in the field laboratory in order to control variation in color readings. For instance, Sam constructed a cardboard box closed on three sides by white walls and installed a daylight incandescent lamp. In anthropological and geological fieldwork samples are investigated directly in the field (e.g., Goodwin, 1996; Latour, 1993a). In the move to conduct color determination in the laboratory and hence in an even more isolated and decontextualized situation, we observed considerable changes in the practices associated with the Munsell chart from those usually observed.

Sam's double move from the field into the laboratory and then into the confines of the cardboard set-up followed by an additional cleaning of the animals literally abstracts (< Lat. *ab-*, away, + *trahere*, to draw) the color. Whether this has any relevance for the color in the field has not been determined. But the practice of decontextualizing the lizard was an attempt at satisfying the scientific community's norm of repeatability, consistency, and the reduction of observer bias that can be attributed to evidence and therefore contribute to a strengthening of the "facts" reported. (These are modalities that other researchers in the community can use to make counter claims.) These practices are also consistent with the requirements of extended statistical analysis in which individual lizards and groups are compared and processed *ceteris paribus*. Comparing lizards on the basis of data collected in a special place irrespective of their native home range is the scientific practice related to the implementation of *ceteris paribus* conditions.

Sam was concerned with accountability—that is, she had to be able to account for her practices in such a way that others, who could potentially question her practices, would not modalize her claims. However, despite the multiple moves in decontextualization, Sam was concerned that her color measures may not be "repeatable" and "consistent" enough or without "observer bias on color measure." To deal with the variability, and to counter any modalizing of her color measure, she needed an index for the amount of variability, she needed "information about the precision of the measure . . . something like standard error." During our presence in the field, Sam did not know whether she could convert the Munsell classifications into numerical data, which she could then easily treat using familiar statistical practices (she had been a teaching assistant in a research design course in a fourth-year undergraduate course for several semesters). Sam decided to enact a "blind" test involving two field assistants. These tests showed that there was agreement with the first and second field assistant on only 2 and 7 of 11 color categorizations, respectively. The agreements on mottle determination were higher, but not good enough for Sam. She therefore decided that she would do the Munsell chart work herself. After running another blind test to get a sense of her own consistency, Sam was discouraged: "I am not sure I am measuring anything. I know I am measuring something, but I don't know if it really means anything in terms of whether it is repeatable."

Shoring up the Evidence

Although nobody holds our scientist accountable then and there,[2] and although it is unlikely anyone would ever enact a data audit, Sam worked hard to make precision (and with it the objectivity of her observations) an accountable aspect of her work. Sam, as the

---

2. Sam is organizationally accountable to her thesis committee, the audience at a conference, or readers of a scientific journal. Members of these groups specify (heterogeneous) constraints on what counts as evidence, what is appropriate replication (repetition), what are acceptable errors, and so on.

other ecologists in the camp, treated "replication" as "repetition of process" which allowed her to check reproducibility of evidence. Enacting repetitive measurement, and therefore being able to draw on it as a discursive resource, contributes to the "authority" of the data and to the credibility that the research is replicable across sites. Replication provided an index of variability. That is, variability is not a problem in itself as long as the research can account for it and, in Sam's case, express it in numerical terms. Ultimately, "replication" becomes the discursive resource for supporting claims about the soundness of the research done and the quality of the data presented.

Some readers may think that replication is a standard practice enacted by scientists. Our ecologists also made claims in this sense: "Doing things repeatedly is a habit." Yet this was not the case across their own practices (e.g., Roth & Bowen, 1999b). When measurement was considered to be unproblematic, such as in the measuring of lizard length or weight, Sam never replicated her measures. On the other hand, "problematic" measurements such as color determination were always accompanied by replication practices. In the present case, the objective character of color descriptions that entered the database emerged from the associated practices that make them rationally accountable. Here, "repeatability," "consistency," and "observer bias" were enacted in the form of independent assessments. The objectivity of the work of measuring and coding was provided for by arrangements that encourage the emergence of an accountable practice. Thus, measurements in themselves are not sufficient as evidence for a claim, but the measurement practices themselves have to be supported by evidence of another set of practices.

Claims and Evidence in a Grade 8 Ecology Course

In this episode, we exemplify what we learned about the construction of claims and evidence in a grade 8 ecology class (13–14 years old) that was taught according to "authentic science" and "cognitive apprenticeship" metaphors.[3] These metaphors encourage teachers to set up learning environments such that students enact practices that have a structural likeness with those that scientists enact (e.g., Brown, Collins, & Duguid, 1989). At the same time, teaching according to these metaphors, if it is not accompanied by a reflective component in the science courses, leads to indoctrination into the particular form of knowing characteristic for Western science (Roth, 2001).

Claim and evidence construction became central to the activities in two classes of a grade 8 ecology course taught by one of us (Bowen), and recorded by the other (Roth) using video and audio recordings, interviews, and artifacts produced by students and teacher. In this course, Bowen used an apprenticeship metaphor to plan the curriculum and to interact with the students, who themselves were asked to use the metaphor "scientist" as a referent in their activities. Throughout the ecology unit, students designed and carried out their own investigations. There were two fundamental requirements that framed the open-ended inquiries. First, students had to provide rationales for all aspects of their research: Why should my research question be of general interest? Why did I operationalize variables in this way? Why are the representations I chose (table, graph, average, etc.) the most suitable ones to make my point? and How do the data support my claims? Second, these rationales had to be regarded as convincing by their peers. Every now and then, the teacher asked students to assemble into discussion groups in which their normal partners could not also be present. In these groups, students presented results of their research, discussed and critiqued each other's research design and claims, and came to understand similarities and differences between their respective investigations and results.

As part of the overall study, we wanted to understand how certain problems, constructed by individual students, would be solved by others. In this case, we constructed paper and pencil versions of what we had observed on videotape as part of our

---

3. An extensive description of the school, the conditions of teaching science in this grade 8 class, and the artifacts constituting our database have been provided elsewhere (Roth, 1995).

ongoing analysis. (Thus, we nested quantitative studies within an ethnographic study. On the particular strength of these nested designs, see Lave [1988] or Brown [1992].) Sometimes we constructed but one form, sometimes alternate forms for investigating "problem solving" in a controlled experimental design. The focus of students' discussion of claims and evidence in the following episode arose from one of these problems.

The Point of Argument

This task arose from the work of Jamie and Mike who had investigated a number of research questions related to the growth of May apples, including light intensity, soil pH, and soil moisture. In one case, the number of plants in one of the reference plots had increased in the days between site visits. They had concluded, and were puzzled by the fact, that the average height of the plants had decreased rather than increased. We therefore constructed a task to be completed by students as homework (Figure 1), and later discussed in groups of three.

The task is interesting in that there are multiple answers that could be provided. For example, if students decided to disregard the new plants, they would find that the amount of growth (i.e., change in height) was the same in the two areas but that the average height was greater in Area 1. If students calculated the average heights of plants from the first sampling period to the second, there was the unexpected outcome of a decrease in Area 2. If students considered total growth in both areas including the new plants, the growth in Area 2 (20 units) was far greater than that in Area 1 (3 units). There were differences in the pH levels and light intensity as well, favoring in one instance a positive relationship with growth, in the other a negative relationship.

The Episode

Students were asked to sit in groups of three with students other than their research partners. They were asked to come to a group consensus about the answers to the questions in their homework assignment, including the rationale(s) for their answers. We video-taped Mike, Shaun, and Fab as they discussed the assignment.

Mike and Shaun each read out their written answers to the homework task. Fab, who had not completed the task, was busy

The Lost Notebook

The following is a problem similar to that which some of your classmates faced when they were working in their ecozone. In your own group, try to solve the problem of this group.

On Monday, Day 1 of your study you counted 3 May apples, a short plant with a single stalk, in each of your two test areas. The height of each plant is recorded in Table 1. On Thursday, Day 4 of your study, you went to your ecozone again and found 3 May apples in Area 1, but 5 May apples in Area 2. The height of each plant is recorded in Table 1. In Area 1 you have 400 foot-candles of light; the pH level was 6.5. In Area 2 the level of light was 300 foot-candles; the pH level was 6.8.

Table 1

Day	Plant #	Area 1	Area 2
1	1	17	19
1	2	20	22
1	3	23	16
2	1	18	20
2	2	21	23
2	3	24	17
2	4		5
2	5		12

Answer the following questions:
a. How much growth do you observe? (Give a reason)
b. In which area do you observe more growth? Why? (Give reasons for your answers)
c. What is the average growth in each area? (Give a reasoned answer)
d. From the information provided, what reasons for the growth patterns would you propose? Why?
e. How could you test your hypotheses?

Figure 1. An "authentic" homework problem that had arisen from the work of grade 8 students. The teachers generated the problem by describing the results one student group videotaped as part of the study. Its intent was to ascertain how other students might deal with a problem of their peers; this form was assigned as homework to all students who then discussed it during the subsequent class.

reading the assignment and did not, at this stage, contribute to the conversation.[4]

1	Mike:	READS[(task) "Question 2, in which area do you observe
2		more growth? Why? Give reasons for your answers."]
3		READS["I observed the most growth in Area 2 because of
4		three days of growing the plants had the largest total unit
5		increase which was 20. Area 1 only had 3 units of in-
		crease."]
6	Shaun:	READS[(from notebook) "Area 2 I observed more growth.
7		This is so because the pH level is closer to neutral which is
8		7. The pH level is 6.8 as was one of my focus questions. I
9		found that if the pH level is closer to 7, the height."]
10	Mike:	It's what each area, what [you *observed*.]
11	Shaun:	[Give reasons,] give reasons for
12		your answers, so I am giving you the reasons.
13	Mike:	You're supposed to do it by the *measure*ments.
14	Shaun:	These are good reasons. Doesn't it say give reasons?
15	Mike:	Yeah.
16	Shaun:	You didn't give reasons, did you?
17	Mike:	Yes I did!

Mike's text suggested that he observed more growth in the second area "because of three days growing." Shaun also suggested that there was more growth in the second area (line 6) but related it to pH levels. He further commented that the pH level in the task statement was the same that he had measured in a similar investigation in his own plot (line 8). Although Mike's next statement was incomplete, he stressed that "it" has to be "what you *observed*" (line 10). If the utterance is heard as a challenge to what Shaun had said, Mike seems to question that Shaun's statement is

---

4. The following transcription conventions are used:

READS["xxx"] Individual reads from problem or written answer;

M: [It's what each]

S: [I said   ] Square brackets indicate overlapping utterances;

= "Latching" so that there is not the conventional temporal space between two speakers;

italic Italics are used to indicate emphasis in speech;

(2.0) time between utterances, in seconds;

.,?! Punctuation in the transcript reflects our hearing of the utterance as a completed sentence, clause, question, or exclamation.

not based on observation. Thus, he can be heard as challenging Shaun's warrant, perhaps suggesting that it was not based on some *observation*. Shaun heard Mike's comment (line 10) as just such a challenge and began by stating the problem he addressed, "Give reasons" (lines 11–12). Mike retorted that the warrant has to be "by the measurements" (line 13), whereas Shaun insisted that his warrant, pH, was a good reason (line 14). When Mike agreed to this assessment, Shaun, in turn, challenged Mike's answer: accordingly he did not even have a warrant (line 16). But Mike rejected this challenge.

In this part of the episode, therefore, both students agreed that there is more growth in Area 2. The central question of the debate is the nature of the warrant required for supporting this claim. Shaun proposed pH, whereas Mike appeared to state the time between observations in his answer and further talked about the measurements themselves as being a sufficient warrant (line 13).

18	Shaun:	What were your *reasons*?
19	Mike:	READS[(from notebook) "I observed more growth in
20		Area 2. This was *because*, in three days of growing, these
21		plants had the largest total unit increase which was 20.
22		And Area 1 only had [3 units] of increase."
23	Shaun:	[Ah, OK.]
24	Mike:	That's my reason.
25	Shaun:	You didn't have a large enough time to grow. And a=
26		[very few]
27	Mike:	=Your reason [is that the] pH is more closer [to seven.]
28	Shaun:	[but that] is
29		still a *rea*son, is it not?
30	Mike:	No, because you don't know that if it is closer to seven here
31		that it's gonna grow better. That's not proven in this lab.
32	Shaun:	Yeah, that's what I am saying, as I did in my lab, I am stat-
33		ing that=

In this second part of the episode, Shaun and Mike argue what would constitute an appropriate reason for each of their answers. Having already agreed on what the growth was, the question here then became the validity of the warrants they respectively provided. In the light of what Mike said earlier (line 13), his utterance in line 19 emphasized that the support for his answer lies in the measurement, or, more accurately, in the difference between the measurements over the three days. But Shaun did not hear the utterance in that way. Rather, what was salient to him was the length of time between the two days on which data were collected. He therefore suggested that the three intervening days were insufficient to explain the amount of growth in Area 2 (line 25). Mike, on the other hand, worked out his own understanding of what Shaun's warrant was, the differences in soil acidity (pH) (line 27). Shaun understood Mike's statement as questioning his own warrant and retorted that it was still unquestioned as a reason for the growth (line 28–29). However, Mike questioned the warrant. He suggested that the data collected in the investigation did not provide evidence for claiming a relationship between growth and pH. That is, he raised doubt by appealing to the authority of experimentation needed to be able to prove a fact, which is not in the case they discussed here. But Shaun drew on the authority of his own experiment that he had earlier conducted to answer a research question about the relationship between soil pH and growth. Thus, he argued that the present case is consistent with his own findings; he drew on his own experience and investigations as a warrant to support his claim in this problem.

In Shaun's case, the problem he had investigated in the field was not independent of the word problem he and Mike were working on. That is, he drew on his field research experience as a way of understanding the textual and graphical representations. This action is not unlike those we observed used in our ethnographic work among ecologists and in our parallel study of graph interpretations scientists provided. In each case, scientists heavily relied on their embodied understanding of what they understood the representation (graph) as standing for. If they could not draw on such understanding, they failed to provide standard, agreed-upon interpretations of graphs. Here, the warrants that support an interpretation arise from a deep understanding of what the world is like rather than from some abstract manipulation of logic symbols.

Mike questioned Shaun's approach in a more serious way. In fact, he questioned the embodied experience Shaun had brought in support of his answer and suggested that it was of the realm of "thoughts" rather than that of "facts" on which he was to base his answer.

34	Mike:	=You are supposed to be stating facts not thoughts.
35	Shaun:	No, see look (2.0) Give a reasoned answer, that is a fact.
36	Mike:	Give reasons for your answers. *Why*
		[do you observe more?]
37	Shaun:	[Yeah, well, you can see,]
38		[if you can see, right, see . . .]
39	Mike:	[Growth in Area 2? Why] do you, why do you think?
40		[There might be...]
41	Shaun:	[Yeah, exactly, *that*] is why I give you the pH level, right
42		and comparing. They can be all right, that can be too, true
43		too, and 20 units of growth.

Shaun reiterated that he was providing what the question asked him to, "a reasoned answer" (line 35). Mike first restated the sentence from the task, and then framed what he understood to be the "real" question, "*Why* do you observe more?" (line 36). Here, then, Shaun and Mike talked at the same time. Shaun began to stumble under the barrage of "why's." In a last attempt, he extended what he heard as a suggestion that "there might be . . ." other factors as he reiterated that is just why he suggested pH as an appropriate warrant. At this point, he began to acquiesce and suggested that both warrants could be accepted as "true" (line 42).

Warrants and Arguments

In this episode, the question is not only what the nature of the appropriate warrant is but also what exactly has to be warranted, or what the question asks them to warrant. (Recall that the students were enabled by the teacher to research anything they wanted as long as they made convincing arguments for each research study and its results.) Mike can be understood as wanting to provide a warrant for the question "How much growth is there?" whereas Shaun provided a warrant for "Why is there more growth in Area 2?" However, these differences do not come to the fore in the present situation. The question statement allows both interpretations.

Mike answers the first part of question b (Figure 1), whereas Shaun answered what he read as the second part of the question. In science, the warrants for a claim reside in the experimental configuration. Thus, when the equipment is very expensive and only few research groups can afford it, it becomes very difficult to question the claims (Latour, 1987). The more a set of practices, tools, and instruments are black boxed in a more complex instrument, the less likely it is that the experimental results coming from the complex instrument will be questioned.

We can observe another similarity of this discussion with scientific argumentation in the argumentative and *oppositive* mode (Amann & Knorr-Cetina, 1990). Scientists use such oppositive devices to work out potential problems with the evidence that they previously constructed; sometimes this is enacted along the metaphor of "playing the devil's advocate." Here, the overlapping utterances (lines 10–11, 22–23, 26–27, 27–28, 36–37, 38–39, and 40–41) and the combative tone of the interaction had a very similar quality to those oppositive discussions of scientists. In this class, students were not rewarded for "right answers" but for providing sufficient evidence that, in the first instance, could convince their peers rather than the teacher. Thus, we see grade 8 students enact a practice of working out what can count as evidence.

Evidence for Epistemological Claims in High School Physics

In this section, we exemplify what can happen when teachers involve their students not only in learning science but also in reflecting on the form that scientific (and common sense!) claims and evidence are constructed. Most students do not have opportunities to learn and enact epistemological discourse as part of their school experience (Désautels & Roth, 1999). We may therefore not be surprised that students seem to have limited discursive repertoires for talking about epistemology, nature of science, and similar topics. However, our own research shows that students develop quite sophisticated ways of talking epistemology in the context of science (e.g., Roth & Alexander, 1997; Roth & Lucas, 1997; Roth & Roychoudhury, 1994). In this section, we present an excerpt from an extensive study of students' epistemological discourse and change over an 18-month period. During this period, a group of students ($N = 25$) moved from attending grade 12 to grade 13 physics. The database accumulated consists of over 3,000 typed pages of transcribed interviews, class discussions, and student essays in which students talk and write about epistemology in the context of science. The teacher (Roth) had added to the formal physics curriculum an epistemological component, which required students to read chapters, articles, and books dealing with various aspects of objectivity, human experience, and representation of knowledge (e.g., physics as language). Our analyses showed that students developed an epistemological practice rather than confronting science or epistemology as (monumental) bodies of knowledge. As they enacted this practice, the problem of what constitutes evidence for scientific concepts was raised over and over again.

Are Magnetic Fields Real?

As part of their discussions, students drew on the available science equipment to enact an investigation that they could then discuss in terms of the epistemological issues. Thus, because students also studied physics, they were provided with multiple opportunities: to question (a) the heretofore unquestioned nature of scientific knowledge and (b) the relation of this knowledge to the evidence that is said to support it. Such questioning is exemplified in the following discussion. The discussion emerged from a question whether scientific concepts are real, and what evidence we have to support the existence of phenomena and concepts.

As this issue was raised and as Mark began to talk about magnetic fields (line 1), Todd turned around and pulled a box with magnets and iron filings from the cupboard. He placed a sheet of paper on top of a magnet and sprinkled it with iron filings that lined up as shown in Figure 2.

1	Mark:	Magnetic fields are real. You can take a magnet, put a
2		sheet over it and pour iron filings on it, and you can ac-
3		tually see magnetic fields.

Figure 2. Iron filings sprinkled over a horseshoe magnet provide the point of departure for a discussion of the question, "Are magnetic fields real?"

4	James:	Not necessarily.
5	Teacher:	What *do* we see?
6	James:	We see the iron filings, we don't necessarily see the fields.
7	Todd:	We see a pattern. It is obvious . . .
8	James:	You don't. It could be anything, it doesn't have to be ne-
9		cessarily what we think.

As was usual in this class, a whole class discussion followed students' reading of two chapters from *Inventing Reality: Physics as Language* (Gregory, 1990). The teacher had started the conversation by asking students about the ontology of concepts, particularly those of "energy" and "field." The students moved their seats so that they could all see the semicircular pattern around the ends of the horseshoe magnet that Todd had prepared. The discussion began with a perceptual phenomenon. All students agreed that they could see some pattern, but the question was raised whether this pattern is sufficient to support the claim that "fields" exist. Mark claimed that what they saw was evidence for fields. But possibly encouraged and scaffolded by the teacher's question (line 5) that encouraged a focus on the nature of the sense data—James and Todd suggested that the visual pattern alone did not constitute evidence for the reality of "fields."

Here, beginning with the teacher's question as a seed and their own preparation of a shared visual display, these students were enabled to question whether sense data constitute sufficient "evidence" for a scientific concept. At this point, the three students who had talked thus far stated not only their preference, but also, in so far as their preferences were dissimilar, that they disagreed about the issue. Mark drew on the empirical (perceptual) repertoire to shore up his claim ("you can actually see"). Here, then, Peter made a decisive discursive move that gave the conversation a new twist. (Several intervening comments have been dropped because they do not contribute to the present discussion.)

10	Peter:	I am just making it up. You exist, because I just made a major
11		statement. Maybe I am all alone, you see? And to stop my
12		loneliness, to stop me from going crazy I invented all of you
13		guys, including physics class.

After James' comment that "it could be anything" that brought about the visual experience of a pattern, Peter raised the stakes by putting forward a radical proposition. Peter suggested that sense data, and with it reality as we experience it, exist because we construct them in our language. To those readers who have followed the discussions surrounding radical constructivism (e.g., von Glasersfeld, 1989), such public discursive moves by opponents of constructivism are probably not unfamiliar. However, here, a grade 13 student produced the statement. (Elsewhere we provide some evidence of the similarities between the epistemological discourse of these students and that of von Glasersfeld and other radical constructivists [e.g., Désautels & Roth, 1999; Roth & Alexander, 1997].) This changed the conversational topic from the relation of sense data and evidence to one of the relation between reality and language (i.e., representation). Peter articulated the epistemological question whether language does in fact create sense data.

The conversation then took an excursion to the nature of electromagnetic waves during which both Todd insisted that these waves exist "because we defined them." James added that "we may not know exactly what it is, but we called it energy, so therefore it exists" and Tom said that "[fields are] something that helps us explain." James then continued the conversation related to the magnetic fields in stating that in the sense of definition and helping to explain, fields existed.

14	James:	So in that sense, in that sense, fields exist.
15	Tom:	But you can't prove it.
16	James:	But we've called that little circle around the iron filings
17		around the magnet, we called that field, whatever that does
18		that, we called it a field, but we don't know what it is.
19	Tom:	Fields exist, but we might not . . .
20	Todd:	But it *is* a field, because . . .
21	Craig:	It could be something completely different, right?
22	Tom:	It is a field, not? Since human beings have named it? It *is* a
23		field therefore it is a *field*. It *is* a field, so we *named* it a
		field.
24	Todd:	No it doesn't exist *because* we named it. It does exist, but *we*
25		have given it a term.

That is, James suggested that fields existed in as far as they explained the sense data, it is a name for the observed pattern; he is supported by Tom's statement that a proof could not be provided (line 15). James then elaborated the discourse about the relationship between sense data and discourse (lines 16–18). Accordingly, there are some visual patterns to which the label field is assigned, but we, as knowing subjects, did not have direct access to whatever created the field. James, is therefore not far from the phenomenological position that naming and recognizing are aspects of the same process (e.g., Merleau-Ponty, 1964; Lyotard, 1991). That is, manifestations of magnetic fields (sense data, pattern in the iron filings) and intuitions of and about magnetic fields (whatever the causal agent) are indistinguishable. Tom (lines 22–23) again raised the issues of whether the naming brings about the field or whether the field exists and humans therefore named it. Todd suggested that "it" existed, and the naming was enacted a posteriori (lines 24–25).

World and Representation

Thus, after Peter's turn (lines 10–13), the conversation took a decisive turn in that the topic shifted from the evidence to language as representational medium. As the consecutive discourse contributions elaborated and modified each other, language (as the principal medium of representation) and its relation to sense data as evidence for scientific concepts became the central issue. Here, then, these students talked and learned to talk not only science (magnetism), but more importantly (because few would go on to study science but went into more lucrative fields of business, medicine, etc.) came to question the role of representations such as language to knowledge. That is, they interrogated the nature of scientific evidence itself rather than naively accepting sense data as authority. Because this conversation happened in the context of a physics class, students thereby developed both discursive (and material) practices of physics, and, epistemological discourse that allowed them to take a reflexive stand to knowledge more generally. That is, these students began to interrogate not only scientific knowledge, but more importantly their everyday common sense perceptions and understandings. Easily accessible during the conversations, the equipment at hand afforded students with the construction of objects they could take as shared or over which they could negotiate just what the shared object was. The presence of the materials enabled the discussion to shift between the two forms of discourse, each interrogating the other in terms of the suitability to help us understand nature and knowing about nature.

Science and Epistemology: For a Democratic Society

In this chapter, we present exemplary episodes from three quite different research studies that we conducted in the past. Having conducted all three studies, we are in a unique position to make comparisons across interpretive studies and thereby test the "transportability" of findings across context (the qualitative/ethnographic researcher's equivalent to generalizability). We now expand our discussion, boldly and perhaps dramatically so, by calling for a science education that makes a contribution to a more democratic (and therefore more free) society at large.

Our episodes exemplified that scientists' epistemological issues were internal to their practice in the sense that they did everything possible and thinkable to shore up their evidence in order to make it unassailable. The students, on the other hand, also engaged in discussions of the nature of the evidence itself. In the case of the grade 8 students, the argument was more similar to that of the scientists because the relationship between evidence and claims was internal to science. In the case of the physics students, the discourse was fully epistemological in that they discussed the relation that evidence has to sense data, and therefore questioning scientific knowledge from the outside. We suggest that in a democratic society people should be in a position to evaluate claims that are made by all the stakeholders in a particular issue. It is only if knowledge claims from a variety of sources are evaluated as to their suitability to contribute to the issues at hand that we can make informed decisions. This pertains not only to scientific knowledge, but knowledge of various sources including common sense, culturally specific knowledge (e.g., traditional ecological knowledge), or spiritual knowledge.

Such an approach is feasible as can be seen, to take just one example, in the recent developments surrounding the research and treatment of AIDS. Initially, scientists used their own domain-internal criteria for establishing research and treatment protocols. AIDS activists made it clear that this approach was dehumanizing and less than fully informed. Over time, they understood how to enter the debate with the consequence that new research and treatment protocols were elaborated that also accounted for criteria external to science. We argue that the notion of a democratic society implies that the discourses are open, and single-domain expertise is rarely enough to answer the difficult questions society faces. The Scandinavian model of participatory design of workplaces is one example where the traditional reliance on single-domain expertise has been overcome to the benefit of all participants leading to workplaces better adapted to the needs of the worker-users (e.g., Ehn, 1992).

By including epistemology as part of our educational project, we enabled students to co-participate in conversations during which students relativized forms of knowledge and knowledge production. Science education is in a particular position to educate students to be critical, because it has been in the past quite dogmatic itself. Because co-participation leads to co-learning (e.g., Roth, 1998), these conversations (and the associated readings and essays) afford students opportunities to evolve their discursive practices relative to epistemological issues. In addition, encouraging students to engage in discussions of this sort helps them to learn to reflectively critique their methods, knowledge constructions, and claims from their investigations in the context of being in a public forum. Something similar occurred when Sam critiqued her own work in light of anticipated, evaluative conventions of the wider scientific community. However, the implications of this pedagogical approach extend considerably beyond the walls of the classroom and the halls of the school.

Our research on graphing suggests that students often accept a scientific claim prima facie as "hard" facts. They do so as if there was a one to one correspondence between the claim and an observation. They ignore the ladder of facticity and the conceptual leaps that occur as a claim was established and provide the claim with a degree of authority that is perhaps unwarranted. Engaging students in this type of discourse, in which they interrogate the foundations of claims made by others, helps develop their understanding that theory, raw measurement, evidence, and claims are mutually constitutive and that they are not necessarily clear cut in what they immediately represent or (come to) stand for. Again, science education can make a unique contribution because facts, theory, evidence, claims and so forth are articulated and taken for granted much more so than in other domains. Our experience with the grade 12 and 13 physics students showed that once they experience scientific "facts" as fallible, they are more than likely to interrogate other forms of knowledge in an equally critical way. Our examples of students using data to support theory and theory to support data suggest that they begin to better understand that theory does not stand on own but can also provide support for the very evidence which is in turn supporting the theory. This is the beginning of an informed, reflective citizenry.

As a society, we can only become free from indoctrination if a sufficient number of members are enabled to interrogate all forms of knowledge and knowledge construction (e.g., Feyerabend, 1975, 1978). In such a society, the attention is switched from the demands of the object, evidence-claim relationship internal to science, to the demands of the purpose for which a particular inquiry serves (Rorty, 1991). In order to free ourselves from any doctrine (scientific, religious, or common sense), we need to be able to create a hall of mirrors effect which gets us to stop what nature really is and to settle for an ever-expanding choice of representations. Such ability would then allow people to interrogate claims and evidence put forward in a variety of public arenas. For example, scientific megaprojects such as the Human Genome Project were never put to formal votes of the general population or the scientific community. Rather, the US Congress funded megaprojects on the basis of elite and self-serving scientific testimony from those considered "knowledgeable" about the benefits—who were also insiders. On the other hand, there is an increasing trend to involve ordinary citizens in participatory decision making about scientific and technological issues as they pertain to society (e.g., Rowe & Frewer, 2000). For example, in Switzerland, the entire population was invited to contribute, through their referendum votes, to the Swiss decision-making with respect to biotechnology. We might expect different outcomes of the discussions surrounding megaprojects if a larger proportion of the population (and with it, a larger proportion of the politicians) could have questioned the testimony about the benefits of these projects. Yet it is possible for citizens to successfully participate with scientists on panels, contributing to the learning of all as was shown in the recent "Citizens' Panel on Telecommunication on the Future of Democracy" held in the US (Guston, 1999). To prepare students to be members (of a democratic society) who have the competence to enact such interrogations, we, as science teachers, must allow students to learn not only how to use the various discourses as resources, but ways of critically questioning these discourses. In the way a democratic society scorns any form of religious indoctrination (see how fundamentalist movements in countries such as Iran or Afghanistan are regarded in the West), we should also scorn indoctrination into scientific discourse, or, even more insidiously, common sense discourse. But because we can only seriously and effectively interrogate what we know and are familiar with, students should be enabled to both learn scientific discourse and interrogate the very same discourse.

Our ultimate call therefore is that science education should not contend itself with teaching students science, or even how to construct evidence (e.g., as in our grade 8 case) so that it becomes more unassailable and therefore better supports the claims they make. Rather, as our experience in the grade 12 physics course shows, students need to be able to engage in and develop their epistemological discourse practices and take a more reflective stance to all forms of knowledge, scientific or otherwise. Interrogating knowledge and knowledge construction, of course, is at the heart of border pedagogy, itself concerned with educating students for life in an open and free society (e.g., Giroux, 1992). Although we remain science teachers and science educators at heart, we are also concerned with the larger issues of *cui bono*? Who is science for? What purposes does scientific knowledge serve? Who benefits from unquestionable scientific knowledge? We therefore have to ask, "What is it about science that can help us to lead science education away from the reproduction of inequalities in participation in science along the lines of gender, race, social class, and expertise?" (How does "scientific" testimony come to outweigh other testimony in courts, construction of "learning disabilities," etc.?)

We side with the calls for an ideal democratic (and therefore free) community, in which reigns a sense of solidarity. Such a community would practice a philosophy-of-wisdom inquiry in which the discourses of music, literature, drama, politics, science, religion, and philosophy are treated at the same level (Maxwell, 1992). However, absolutist discourses that cannot be interrogated (scientific, religious, or otherwise) would not have a place. Rather, all forms of discourses would be open to interrogation, subject to a social epistemological inquiry of recontextualizing, which no longer requires the labels "science" and "scientific" for taxonomic or authoritative purpose. Ultimately then, our students and future citizens would then be enabled to both draw on and question different discourses when it comes to solve the serious challenges to humanity such as AIDS-HIV, the greenhouse effect, food shortages, and so forth.

Acknowledgments

This work was made possible in part by the Social Sciences and Humanities Research Council of Canada in the form of grants 410-96-0681 and 410-99-0021 (to Roth) and a doctoral fellowship from the Social Sciences and Humanities Research Council of Canada (Bowen).

References

Amann, K., & Knorr-Cetina, K. D. (1990). The fixation of (visual) evidence. In M. Lynch & S. Woolgar (Eds.), *Representation in scientific practice* (pp. 85–121). Cambridge, MA: MIT Press.

Bowen, G. M., Roth, W.-M., & McGinn, M. K. (1999). Interpretations of graphs by university biology students and practicing scientists: Towards a social practice view of scientific representation practices. *Journal of Research in Science Teaching, 36*, 1020–1043.

Brown, A. L. (1992). Design experiments: Theoretical and methodological challenges in creating complex interventions in classroom settings. *The Journal of the Learning Sciences, 2*, 141–178.

Brown, J. S., Collins, A., & Duguid, P. (1989). Situated cognition and the culture of learning. *Educational Researcher, 18*(1), 32–42.

Derrida, J. (1988). *Limited inc.* Chicago: University of Chicago Press.

Désautels, J., & Roth, W.-M. (1999). Demystifying epistemology. *Cybernetics & Human Knowing, 6*(1), 33–45.

Edwards, D., & Potter, J. (1992). *Discursive psychology*. London: Sage.

Ehn, P. (1992). Scandinavian design: On participation and skill. In P. S. Adler & T. A. Winograd (Eds.), *Usability: Turning technologies into tools* (pp. 96–132). New York: Oxford University Press.

Epstein, S. (1997). Activism, drug regulation, and the politics of therapeutic evaluation in the AIDS era: A case study of ddC and the 'Surrogate Markers' debate. *Social Studies of Science, 27,* 691–726.

Feyerabend, P. (1975). *Against method: Outline of an anarchistic theory of knowledge*. London: New Left Books.

Feyerabend, P. (1978). *Science in a free society*. London: New Left Books.

Fuller, S. (1988). *Social epistemology*. Bloomington: Indiana University Press.

Fuller, S. (1997). *Science*. Buckingham, England: Open University Press.

Gilbert, G. N., & Mulkay, M. (1984). *Opening Pandora's box: A sociological analysis of scientists' discourse*. Cambridge, England: Cambridge University Press.

Giroux, H. (1992). *Border crossings: Cultural workers and the politics of education*. New York: Routledge.

Goodwin, C. (1996). Practices of color classification. *Ninchi Kagaku (Cognitive Studies: Bulletin of the Japanese Cognitive Science Society), 3*(2), 62–82.

Gregory, B. (1990). *Inventing reality: Physics as language*. New York: Wiley.

Guba, E., & Lincoln, Y. (1989). *Fourth generation evaluation*. Beverly Hills, CA: Sage.

Guston, D. H. (1999). Evaluating the first U.S. consensus conference: The impact of the citizens' panel on telecommunication and the future of democracy. *Science, Technology, & Human Values, 24,* 451–482.

Harré, R., & Gillett, G. (1994). *The discursive mind*. Thousand Oaks, CA: Sage.

Jayyusi, L. (1991). Values and moral judgement: Communicative practice as a moral order. In G. Button (Ed.), *Ethnomethodol-*

ogy and the human sciences (pp. 227–251). Cambridge, England: Cambridge University Press.

Kuhn, T. S. (1970). *The structure of scientific revolutions* (2nd ed.). Chicago: University of Chicago Press.

Larochelle, M., & Désautels, J. (1992). *Autour de l'idée de science* [Around the idea of science]. Sainte-Foy, Québec: Les Presses de Laval.

Latour, B. (1987). *Science in action: How to follow scientists and engineers through society*. Milton Keynes, England: Open University Press.

Latour, B. (1993a). *La clef de Berlin et autres d'un amateur de sciences* [The key to Berlin and other lessons of a science lover]. Paris: Éditions la Découverte.

Latour, B. (1993b). *We have never been modern*. Cambridge, MA: Harvard University Press.

Latour, B., & Woolgar, S. (1986). *Laboratory life: The social construction of scientific facts*. Princeton, NJ: Princeton University Press.

Lave, J. (1988). *Cognition in practice: Mind, mathematics and culture in everyday life*. Cambridge, England: Cambridge University Press.

Lederman, N. G. (1992). Students' and teachers' conceptions of the nature of science: A review of the research. *Journal of Research in Science Teaching, 29*, 331–359.

Lewenstein, B. V. (1995). From fax to facts: Communication in the cold fusion saga. *Social Studies of Science, 25*, 403–436.

Lucas, K. B., & Roth, W.-M. (1996). The nature of scientific knowledge and student learning: Two longitudinal case studies. *Research in Science Education, 26*, 103–129.

Lyotard, J.-F. (1991). *Phenomenology*. Albany: State University of New York Press.

Maxwell, N. (1992). What kind of inquiry can best help us create a good world? *Science, Technology, & Human Values, 17*, 205–227.

Merleau-Ponty, M. (1964). *L'œil et l'esprit* [The eye and the mind]. Paris: Gallimard.

Mulkay, M., & Gilbert, G. N. (1983). Scientists' theory talk. *Canadian Journal of Sociology, 8*, 179–197.

Nutch, F. (1996). Gadgets, gizmos, and instruments: Science for the tinkering. *Science, Technology, & Human Values, 21,* 214–228.

Pollner, M. (1987). *Mundane reason: Reality in everyday and sociological discourse.* Cambridge, England: Cambridge University Press.

Potter, J., & Wetherell, M. (1987). *Discourse and social psychology: Beyond attitudes and behaviour.* London: Sage Publications.

Quine, W. V. (1995). *From stimulus to science.* Cambridge, MA: Harvard University Press.

Rorty, R. (1989). *Contingency, irony, and solidarity.* Cambridge, England: Cambridge University Press.

Rorty, R. (1991). *Objectivity, relativism, and truth: Philosophical papers* (Vol. 1). Cambridge, England: Cambridge University Press.

Roth, W.-M. (1995). *Authentic school science: Knowing and learning in open-inquiry laboratories.* Dordrecht, Netherlands: Kluwer Academic Publishing.

Roth, W.-M. (1998). Science teaching as knowledgeability: a case study of knowing and learning during coteaching. *Science Education, 82,* 357–377.

Roth, W.-M. (2001). 'Authentic science': Enculturation into the conceptual blind spots of a discipline. *British Educational Research Journal, 27*(1), 5–27.

Roth, W.-M., & Alexander, T. (1997). The interaction of students' scientific and religious discourses: Two case studies. *International Journal of Science Education, 19,* 125–146.

Roth, W.-M., & Bowen, G. M. (1999a, August). Are graphs really worth ten thousand words? Paper presented at the biannual meeting of the European Association for Research on Learning and Instruction, Sweden.

Roth, W.-M., & Bowen, G. M. (1999b). Digitizing lizards or the topology of vision in ecological fieldwork. *Social Studies of Science, 29,* 719–764.

Roth, W.-M., & Bowen, G. M. (1999c). Of cannibals, missionaries, and converts: graphing competencies from grade 8 to professional science inside (classrooms) and outside (field/laboratory). *Science, Technology, & Human Values, 24,* 179–212.

Roth, W.-M., & Bowen, G. M. (in press). Of disciplined minds and disciplined bodies. *Qualitative Sociology.*

Roth, W.-M., & Lucas, K. B. (1997). From "truth" to "invented reality": A discourse analysis of high school physics students' talk about scientific knowledge. *Journal of Research in Science Teaching, 34,* 145–179.

Roth, W.-M., Masciotra, D., & Bowen, G. M. (in press). From thing to sign and object': Toward a genetic phenomenology of graph interpretation. *Science, Technology, & Human Values.*

Roth, W.-M., & McGinn, M. K. (1997). Science in schools and everywhere else: what science educators should know about science and technology studies. *Studies in Science Education, 29,* 1–44.

Roth, W.-M., McRobbie, C., & Lucas, K. B. (1998). Four dialogues and metalogues about the nature of science. *Research in Science Education, 28,* 107–118.

Roth, W.-M., McRobbie, C., Lucas, K. B., & Boutonné, S. (1997). The local production of order in traditional science laboratories: A phenomenological analysis. *Learning and Instruction, 7,* 107–136.

Roth, W.-M., & Roychoudhury, A. (1994). Physics students' epistemologies and views about knowing and learning. *Journal of Research in Science Teaching, 31,* 5–30.

Roth, W.-M., & Tobin, K. (1996). Aristotle and natural observation versus Galileo and scientific experiment: An analysis of lectures in physics for elementary teachers in terms of discourse and inscriptions. *Journal of Research in Science Teaching, 33,* 135–157.

Rowe, G., & Frewer, L. J. (2000). Public participation methods: A framework for evaluation. *Science, Technology, & Human Values, 25,* 3–29.

Schnabel, J. (1994). Puck in the laboratory: The construction and deconstruction of hoaxlike deception in science. *Science, Technology, & Human Values, 19,* 459–492.

Simon, B. (1999). Undead science: Making sense of cold fusion after the (arti)fact. *Social Studies of Science, 29,* 61–85.

von Glasersfeld, E. (1989). Cognition, construction of knowledge, and teaching. *Synthese, 80,* 121–140.
Wittgenstein, L. (1994/1958). *Philosophical investigations* (3rd ed.). New York: Macmillan.

Appendix

CHAPTER 1: In our introduction, we provide a narrative that integrates the major themes addressed in the various chapters. For readers who want to construct their own narrative by reading the chapters in a sequence of their own, we provide a brief summary of each in the sequential order that these appear in the book.

CHAPTER 2: A number of recent and not-so-recent curriculum initiatives have sought to link science education with action, sometimes under the heading of "citizenship." In his chapter "Linking School Science Education with Action," Edgar Jenkins reviews the origins and nature of some of these initiatives in a variety of contexts. He also examines the implications of seeking to link science education with action for the science curriculum, for science itself, for pedagogy and assessment, and for schooling in general. As his argument unfolds, Jenkins suggests that there are two other factors involved in the relative lack of success of earlier curriculum initiatives focusing on citizenship and practical action. First, Jenkins argues that earlier curriculum initiatives accommodated a partial and under-theorized view of the role that science plays in the modern world. In the past, science was seen as the domain of heroic efforts in which "great" thinkers and researchers cull truth from nature. However, such a description of science is no longer viable given the many forms in which it exists in society today. Second, if such engagement has taken place at all, it has been vicarious rather than personal. Furthermore, such efforts paid insufficient attention to the wider consequences that arise when science education is linked with action. Jenkins discusses both of these issues in greater detail in Parts 2 and 3 of his chapter.

CHAPTER 3: In the chapter "Learning Science (Informally) Through Sociopolitical Action," Lee and Roth take a look at science as it becomes salient in the activities of an environmental activist group. Their research is influenced by the work on everyday cognition that has shown that people appropriate many discursive resources (i.e., they acquire knowledge) as they pursue goals of their interest. In this case study, the authors show how members of different groups within one suburban community (e.g., environmental activists, scientists, farmers, residents) learn as they pursue the goal to preserve and improve on of their local watersheds. In order to achieve their goals, often differing and supported in discursively different ways, the activists also appropriate scientific discourse about ecology including habitats, and also appropriate practices related to habitat restoration (e.g., how to build riffles that increase oxygen levels). When this learning is compared to the type of learning occurring in science classrooms, important differences become evident. We use our case study to make suggestions for designing different classroom instruction.

CHAPTER 4: In "Breaking the Spell: Science Education for a Free Society" (Chapter 4), Roth and Lee suggest that science education has been, for the longest time, under the spell of scientists. Science educators and science teachers have subserviently done the job that scientists wanted them to: reproducing a sorting system and existing inequalities. The authors therefore raise some serious questions such as, "Is this the kind of science we want to teach?," "Do we want to continue to use science education as a career selection mechanism or do we want science for all?," and "What would be an appropriate science for all?" They argue for a science education as/for sociopolitical action that constitutes a further development of arguments they had made earlier. Their detailed case study of grade 7 students, who learn science by collecting data in a local creek that later become part of an exhibition organized by environmental activists, features students who already participate in the activities of their community at large. Particularly the great range of people involved in the science lessons makes the boundaries between school and everyday life much less distinct. These boundaries become transparent, allowing practices to move in and out of school more freely. Science education becomes part of what happens at the community level.

CHAPTER 5: In "Teaching Controversial Science for Social Responsibility: The Case of Food Production," Roger Cross and

Ronald Price are concerned with science education as a medium for developing social responsibility in students. The authors argue that teaching science differently is not merely a question of making science more relevant to those who felt disempowered and excluded in the past. Rather, future scientists themselves need to be better aware of the social issues involved in today's (and more so, tomorrow's) science. While they admit that there are dangers involved when teachers begin to teach science organized around socially relevant topics rather than around traditional subject matter themes, they show how students will not lose out on science while focusing on the big issues of today. They articulate four big issues: nutrition and health, the production of food, a sustainable agriculture, and genetically engineered food production. After a brief discussion of three of these issues (the fourth issue, genetically engineered foods cuts across the other three), they provide in sustainable agriculture a detailed argument for and example of a redesigned curriculum.

CHAPTER 6: Since moving to Philadelphia in 1997, Ken Tobin has had a particular interest in the education of inner-city students. Many students in the school where he does his research are absent for high proportions of the time; those in attendance show little enthusiasm for learning; disruptive and disrespectful behavior is rife and performance levels are very low. In this context, Tobin prepares future science teachers who have as their goal to teach in urban schools. Dissatisfied with the traditional gap between university rhetoric and practice of teaching, he began to teach science himself. Going from the high grounds of academe to the reality of everyday teaching has allowed Tobin to perceive and understand urban science education in new ways. Tobin's chapter therefore provides us with a case study that illustrates how science education as praxis is a context for transforming teaching and our understanding of science teaching as praxis. Thus, science education does not only have transformative potential at the student level, but also provides a context for enacting sociopolitical action at the level of the teacher.

CHAPTER 7: Over the past decade, Glen Aikenhead has shown a particular interest in curricula reforms as these pertain to Aboriginal people. In his chapter "Whose Scientific Knowledge?

The Colonizer and the Colonized," he argues that during science education curriculum reform we need to pay particular attention to differences in the way cultures think about and enact knowledge. From his perspective, Aikenhead detects more clearly how the values implicit in Western science have a colonizing quality, which is tacitly hidden in national reforms in science education. Aikenhead suggests that when Aboriginal people can contribute their own ways of knowing (process) and knowledge about nature (facts, concepts), power relationships tend to change within science education. In the process, the conventional colonizer-colonized hierarchy is replaced by a value of sharing for the sake of survival of all. Aikenhead concludes his chapter with an account of learning science in two worlds, the Western and Aboriginal, for the purpose of global citizenry (global in the sense of respect for all, rather than global in the sense of economic colonization). He provides an account of an ongoing project in which Aboriginal people contribute to the way science curriculum is planned and taught and in which their own ways of knowing and knowledge constitute an integral part of what is taught.

CHAPTER 8: Angie Barton and Margery Osborne are science educators who work with minority and at-risk children. They are particularly interested in questions such as, "What is social action in science?," "What is social action in education?," "Where is science education located and whose purposes does it serve?," and "Where do children and teacher 'fit' into science and science education with/for social action?" In their chapter "Remembrance of Homeplace," they explore these questions by examining the concept of "homeplace" as a construct brought to their science classes by children which recasts the science as something transformative. Barton and Osborne argue from a feminist position that is committed to social change. They demonstrate that fundamental to science and education with/for social action are the critical articulations (acts of "remembrance") of both the sociohistorical lives of the children and teacher involved. These acts of remembrance include the experiences, values and ideas students bring to the science classroom, their "homeplaces," as well as how these sociohistorical lives get positioned with and against science (often experienced as a "harsh world").

CHAPTER 9: Nancy Lawrence and Margaret Eisenhart write about their research among pro-life and pro-choice activists. Although this work had been an integral part of their overall project (which involved other doctoral students of Eisenhart as well), they could not publish it together with their other studies in their *Women's Science* (Eisenhart & Finkel, 1998). Among others, the different reviewers and editors who considered their book may have been afraid to deal with the potentially contentious issue (at least in the US) at the interface between science and religion (Eisenhart, 2000).[1] However, readers of this book will find that we can learn a lot about how scientific discourse becomes a resource for making distinctly different arguments: one set of arguments favors abortion, the other the position opposing it. In this chapter, Lawrence and Eisenhart show that learning science does not end with high school. Rather, their work shows how science education can be considered a life-long pursuit whichever side one takes in the ongoing sociopolitical debates.

CHAPTER 10: Marie Larochelle contends that science education as it is practiced today has a particular unreflexive dimension to it. Rather than asking questions such as, why should we accept this classification over another one, science education has been teaching its facts and concepts as "truths." In her chapter "Science Education as an Exercise in Disciplining versus a Practice of/for Social Empowerment," Larochelle articulates the dangers that arise from unreflective approaches to knowledge and provides the example of the difficulties early scientists experienced in classifying the platypus. She re-analyzes several science classroom examples to show how children were much more reflexive about the problematic nature of knowledge than their teacher, and how the (latent) conflicts are settled by drawing on power differences between student and teacher. In this chapter, Larochelle is ultimately (but implicitly) calling not only for science education to be more reflexive about the status of the knowledge that is taken for

---

1. Because of similar fears, four out of five reviewers of the *Journal of Research in Science Teaching* who were sent "The Interaction of Students' Scientific and Religious Discourses" (Roth & Alexander, 1997) refused to review the article. It was subsequently published in a journal based in Europe.

granted, but education more generally. She calls for children to be authorized and authorizing themselves to ask questions and participate in public discourses surrounding science.

CHAPTER 11: In their chapter, Désautels, Garrison, and Fleury call for an "enactment of epistemological practice as subversive social action, the provocation of power, and anti-modernism." Drawing on an analysis of 10-year-old school children, who argue with great skill the nature of the evidence that can be gained from a photograph, the authors show that enacting epistemological practice is not restricted to philosophers and adults. The authors argue that enacting epistemological practice is a form of social action that serves to subvert the social hierarchies that define not only modern but classical (Greco-Roman) culture. They suggest that allowing students to bring their ways of knowing, values, and beliefs into a living dialogue within the classroom and community is the best road to scientific and technological literacy. Their ultimate goal is an education that is liberating rather than liberal.

CHAPTER 12: In this chapter, Roth and Bowen construct an argument for teaching science not just as a body of knowledge but as a way of engaging students in a dual journey where they learn to talk science *and* epistemology. In their itinerary, Roth and Bowen report from their research among three different groups of people involved in science learning. Thus, the authors present in their three exemplary episodes (a) the construction of claims and evidence in an ecological field camp, (b) a grade 8 class studying ecology, and (c) a grade 12 class studying epistemology within their advanced physics course. Roth and Bowen draw out some of the similarities and differences between professional science and the innovative school science courses. The authors conclude with a call for an approach to science teaching that integrates the subject matter and epistemology as a step toward the greater and more democratic goal of involving citizens from all walks of life in making technoscientific decisions.

References

Eisenhart, M. A., & Finkel, E. (1998). *Women's science: Learning and succeeding from the margins*. Chicago: University of Chicago Press.

Eisenhart, M. A. (2000). Boundaries and selves in the making of science. *Research in Science Education, 30*, 43–56.

Roth, W.-M., & Alexander, T. (1997). The interaction of students' scientific and religious discourses: Two case studies. *International Journal of Science Education, 19*, 125–146.

About the Contributors

GLEN AIKENHEAD is Professor of Education, University of Saskatchewan (glen.aikenhead@usask.ca). He has always embraced a humanistic perspective on science, which was enhanced during his graduate studies at Harvard University. It has guided his research into curriculum policy, student assessment, classroom materials, classroom instruction, and cross-cultural science education. His publications include *Science in Social Issues: Implications for Teaching, Views on Science Technology and Society* (with A. Ryan and R. Fleming), *Logical Reasoning in Science & Technology, STS Education: International Perspectives on Reform* (with J. Solomon), and *Rekindling Traditions: Cross-Cultural Science & Technology Units*.

ANGELA CALABRESE BARTON is an associate professor of science education at Teachers College Columbia University, coordinator of the science education program, and Director of the Urban Science Education Center (acb33@columbia.edu). Her research interests include feminist and critical perspectives in science education and its implications in poor urban settings. She has published *Feminist Science Education* and *Teaching Science in Diverse Settings* (Peter Lang). Some of her publications appear in *Curriculum Inquiry, Journal of Research in Science Teaching, Journal of Teacher Education*, and *Women's Studies Quarterly*.

MICHAEL BOWEN is assistant professor of science education at Lakehead University (gmbowen@compuserve.com). His research interests include understanding the development of competency in scientific practices in both formal and informal settings with participants ranging from elementary students to practicing research scientists. He is also interested in the understandings of nature that arise in scientific and non-scientific settings. Some of his publications appear in *Journal of Research in Science Teaching, Cognition and Instruction, Science, Technology, & Human Values*, and *Social Studies of Science*.

ROGER CROSS is a Senior lecturer at the Department of Science and Mathematics Education, University of Melbourne, Melbourne, Australia (r.cross@edfac.unimelb.edu.au). He researches the field of the social responsibility of science and education for formal schooling and the public. His publications include *Science and the Citizen* (co-edited with P. J Fensham), and *Fallout: Hedley Marston and the British Bomb Tests* (2001). He is currently editing a volume dedicated to applying Peter Fensham's writings in science education.

JACQUES DÉSAUTELS is a professor at Laval University (Jacques.Desautels@fse.ulaval.ca). He works in the field of science education and is mainly interested in the epistemological and social stakes involved in the production of scientific knowledges and their re-production in the school context. He has published numerous articles and books including, in collaboration with Marie Larochelle, *Autour de l'idée de science. Itinéraires cognitifs d'étudiants et d'étudiantes.*

MARGARET EISENHART is Professor of Educational Anthropology and Research Methods at the University of Colorado, Boulder (margaret.eisenhart@colorado.edu). Her research focuses on the cultural worlds and social interactions of U.S. girls and women in and around schools. Much of her recent work has examined girls and women in science and technology. She also has written about the uses of ethnographic methods in educational research. Her publications include *Educated in Romance: Women, Achievement and College Culture* (with D. Holland), *Women's Science: Learning and Succeeding from the Margins* (with E. Finkel and others), and *Designing Classroom Research* (with H. Borko).

STEPHEN FLEURY is Professor of Education at Le Moyne College, Syracuse, New York (fleurysc@mail.lemoyne.edu) and currently Director of the Syracuse Center for Urban Education. The themes of his professional writings and presentations in social studies education, science education, teacher education and educational philosophy emanate from his interest in the epistemological impact of science on society, especially pertaining to education for democracy.

JIM GARRISON is a professor of philosophy of education at Virginia Tech in Blacksburg, Virginia (wesley@vt.edu). His research and teaching interests center on pragmatism and especially the philosophy of John Dewey. His most recent books include *Dewey and Eros* and *Constructivism and Education* co-edited with

Marie Larochelle and Nadine Bednarz. His most recent book is *William James and Education* co-edited with Ronald L. Podeschi and Eric Bredo. He wrote the chapter on Education for the companion volume to *The Collected Works of John Dewey* and was an invited participant at the World Congress of Philosophy in 1998 where he spoke on Dewey's theory of philosophical criticism. Jim is a past-president of the Philosophy of Education Society.

EDGAR W. JENKINS is Emeritus Professor of Science Education Policy at the University of Leeds, UK (e.w.jenkins@ education.leeds.ac.uk). His research is concerned with the social and political history of science education and with science education policy in both the contemporary and the historical context. His books include *From Armstrong to Nuffield, Inarticulate Science?, Learning from Others,* and *Science Education: Policy Practice and Professional Judgement.*

MARIE LAROCHELLE is professor at the Faculty of Education of Université Laval, Québec, and researcher at the Centre interdisciplinaire de recherches sur l'apprentissage et le développement en éducation (CIRADE) at Montréal (marie.larochelle@ fse.ulaval.ca). Her research interests pertain to the ways in which students and science teachers frame the world of scientists and, more particularly, the conflicts, negotiations, and socioethical issues that are part and parcel of it. She published (with J. Désautels) *Autour de l'idée de science: Itinéraires cognitifs d'étudiants et d'étudiantes* and edited (with N. Bednarz and J. Garrison) *Constructivism and Education.*

NANCY LAWRENCE is a research consultant with the Consortium for Policy Research in Education (CPRE) at the University of Pennsylvania (lawrence@tradenet.net). She received her Ph.D. in Educational Foundations and Policy from the University of Colorado at Boulder in 1994. At CPRE, she is involved in evaluating a comprehensive school reform effort in the Philadelphia public schools and working with the Merck Institute for Science Education to increase students, understanding of science by changing teachers, classroom practices. "The Language of Science and the Meaning of Abortion" is based on a chapter from Nancy's dissertation, "The Choice of Language and the Language of

Choice," a study of the construction and negotiation of the word "choice" in the contexts of education and abortion.

STUART LEE is a graduate student working on his Ph.D. with Wolff-Michael Roth in the study of scientific practice in community life (shlee@uvic.ca). He is interested in the transformations that scientific practice and representation undergo as they travel from laboratories into municipal halls, neighborhoods and people's yards. His "How Ditch and Drain Become a Healthy Creek: Representations, Translations and Agency during the Re/Design of a Watershed" was published in Social Studies of Science.

MARGERY D. OSBORNE is an associate professor of science education (m-osbor@uiuc.edu). Her research interests include feminist and poststructural perspectives in science education and its implications in early elementary settings. She is author of *Constructing Knowledge in the Elementary School Science Classroom: Teachers and Students*. Some of her publications appear in *Journal of Research in Science Teaching*, *Journal of Curriculum Studies*, and *Women's Studies Quarterly*.

RONALD PRICE, Emeritus Scholar, La Trobe University, is a Senior Fellow at the Department of Science and Mathematics Education, University of Melbourne, Australia (rprice@access.net.au). His research is in the field of comparative education with special reference to China, Marx and education, and the social responsibility of science and education for formal schooling and the public. His publications include *Teaching Science for Social Responsibility*, *Marx and Education in Late Capitalism*, *Education in Modern China*, and *Marx and Education in Russia and China*.

WOLFF-MICHAEL ROTH is Lansdowne Professor of Applied Cognitive Science at the University of Victoria (mroth@uvic.ca). He studies knowing and learning science and mathematics in a variety of formal and informal settings by applying interdisciplinary methodologies derived from linguistics, science studies, cognitive science, and sociology. His publications include *Authentic School Science, Designing Communities*, and *Models of Science Teacher Preparation* (edited with D. Lavoie), and *Being and Becoming in the Classroom; Re/Constructing Elementary Science* (with K. Tobin and S. Ritchie) and *At the Elbow of Another: Learning to*

Teach by Coteaching (with K. Tobin) were both published by Peter Lang.

KEN TOBIN is Professor of Education in the Graduate School of Education at the University of Pennsylvania (ktobin@gse.upenn.edu). His research interests are focused on the teaching and learning of science in urban schools, which involve mainly African American students living in conditions of poverty. A parallel program of research focuses on coteaching as a way of learning to teach in urban high schools. His recent publications include editing (with B. Fraser) of the *International Handbook of Science Education* and (with P. Taylor and P. Gilmer) *Transforming Undergraduate Science Teaching: Social Constructivist Perspectives*, *Re/Constructing Elementary Science* (with W.-M. Roth and S. Ritchie), *At the Elbow of Another: Learning to Teach by Coteaching* (with W.-M. Roth).

Index

Urban
  science education, 125ff, 144ff;
  transformation, 177

Value(s), 101, 106, 108, 111, 118, 121,
  237ff, 248, 251f, 257f, 260, 264;
  trees, 106
Violence
  symbolic 226

Studies in the Postmodern Theory of Education

General Editors
Joe L. Kincheloe & Shirley R. Steinberg

Counterpoints publishes the most compelling and imaginative books being written in education today. Grounded on the theoretical advances in criticalism, feminism, and postmodernism in the last two decades of the twentieth century, Counterpoints engages the meaning of these innovations in various forms of educational expression. Committed to the proposition that theoretical literature should be accessible to a variety of audiences, the series insists that its authors avoid esoteric and jargonistic languages that transform educational scholarship into an elite discourse for the initiated. Scholarly work matters only to the degree it affects consciousness and practice at multiple sites. Counterpoints' editorial policy is based on these principles and the ability of scholars to break new ground, to open new conversations, to go where educators have never gone before.

For additional information about this series or for the submission of manuscripts, please contact:

Joe L. Kincheloe & Shirley R. Steinberg
c/o Peter Lang Publishing, Inc.
275 Seventh Avenue, 28th floor
New York, New York 10001

To order other books in this series, please contact our Customer Service Department:

(800) 770-LANG (within the U.S.)
(212) 647-7706 (outside the U.S.)
(212) 647-7707 FAX

Or browse online by series:
www.peterlangusa.com